# The Next Catastrophe

-------------------------------------------------------

Reducing Our Vulnerabilities
to Natural, Industrial, and
Terrorist Disasters

Charles Perrow

PRINCETON UNIVERSITY PRESS

PRINCETON AND OXFORD

Published by Princeton University Press, 41 William Street, Princeton, New Jersey 08540
In the United Kingdom: Princeton University Press, 3 Market Place, Woodstock, Oxfordshire OX20 1SY

Library of Congress Cataloging-in-Publication Data
Perrow, Charles.
The next catastrophe : reducing our vulnerabilities to natural, industrial, and terrorist disasters / Charles Perrow.
p.   cm.
Includes bibliographical references and index.
ISBN-13: 978-0-691-12997-6 (cloth : alk. paper)
ISBN-10: 0-691-12997-5 (cloth : alk. paper)
1. Emergency management—United States.   2. Disasters—Government policy—United States.   3. Risk management—United States.   4. Hazard mitigation—United States.   5. Terrorism—United States—Prevention.   6. Infrastructure (Economics)—Security measures—United States.   I. Title
HV551.3.P45 2006
363.34'7—dc22      2006037984

British Library Cataloging-in-Publication Data is available
This book has been composed in Sabon with Helvetica Neue Heavy display.

Printed on acid-free paper. ∞

press.princeton.edu

Printed in the United States of America

10   9   8   7   6   5   4   3   2

# Contents

--------------------------

who has tolerated its demands for four years: Barbara, my great good fortune.

I am grateful for support from a National Science Foundation Small Grant award, SES-0329546, and a research fellowship from the Center for International Security and Cooperation, Stanford University, October 2004 to April 2005.

## Acknowledgments

------------------------

MANY COLLEAGUES HAVE commented over the years on this project and sent me references and documents. Foremost among these is Lee Clarke. Among disaster mavens he is tops, and he has been intimately involved in my project from the beginning. At Stanford, where much of the work was done, Scott Sagan, director of the Center for International Security and Cooperation, and assistant director Lynn Eden created a thoroughly stimulating and comfortable atmosphere for workshops, seminars, and colleagueship. Lynn wins the prize for the most thorough and helpful critique of the entire manuscript. The following scholars at Stanford commented on multiple chapters, sometimes extensively: Dara Cohen, Laura Donohue, Michael May, Alex Montgomery, Patrick Roberts, Jake Shapiro, Dan Taron, and Marc Ventresca. Beyond Stanford, I received comments on individual chapters from Dan Berger, Clark Gellings, Eric Lopatin, Darragh Perrow, Jon Pincus, Andrew Shrank, Martyn Thomas, and Barbara Wareck. I thank the fellow members of a panel on software reliability at the National Research Council for unwittingly helping me to understand the Internet and software failures over the past three years. I have given lectures on this topic over the past three years at so many places I have lost track; the questions and probes were always very helpful. Finally, I would like to dedicate the book to an amazing woman

# PART 1

## Introduction and Natural Disasters

-------------------------------------

# 1 Shrink the Targets

------------------------

DISASTERS FROM NATURAL sources, from industrial and technological sources, and from deliberate sources such as terrorism have all increased in the United States in recent decades, and no diminution is in sight.[1] Weather disturbances are predicted to increase; low-level industrial accidents continue but threaten to intensify and the threat of cyber attacks on our "critical infrastructure" becomes ever more credible; foreign terrorists have not relaxed and we anxiously await another attack. Cataclysmic fantasies proliferate on movie screens and DVDs, and scholars write books with "collapse," "catastrophe," "our final hour," and "worst cases" in their titles.

But we have neglected a fundamental response to the trio of disaster sources. Instead of focusing only on preventing disasters and coping with their aftermath—which we must continue to do—we should *reduce the size of vulnerable targets*. Weapons of mass de-

[1]The evidence for the increase in industrial disasters comes from the Swiss reinsurance firm, the world's largest, Swiss Re. The worldwide figures can be found in its Sigma reports. (Swiss Re 2002) "Man-made disasters" include road and shipping accidents, major fires, and aerospace incidents, and the threshold for qualifying is 20 deaths, or 50 injured, or 2,000 homeless, or $70 billion in losses, or insurances losses ranging from $143 million for shipping, $28 billion for aerospace to $35 billion for the rest. Similar criteria are applied to natural disasters. For man-made disasters in the United States, the period from 1970 to 1992 averaged 7.7;

struction (WMDs) already litter our landscape; terrorists need not sneak them in, and they are more likely to be triggered by natural and industrial disasters than by terrorists. Ninety-ton tank cars of chlorine gas are WMDs that travel daily through our cities; dispersing the deadly gas via a tornado or hurricane, an industrial accident, or a terrorist's suitcase bomb would endanger up to seven million people. New Orleans and its surroundings is, or was, our largest port, but it could have been a target one-third the size of its pre-Katrina population of some 450,000 souls, and much easier to defend and evacuate. Because of the increased concentration of the electric power industry, our vital electric power grid is so poorly managed that sagging power lines hitting tree branches in Ohio plunged the Northeast into darkness for hours and even days in some areas in 2003. The industry has made its grid a better target for a heat spell, a flood, a hurricane, or a few well-placed small bombs. Deconcentrating the industry would uncouple the vulnerabilities and barely decrease efficiency, as we shall see.

Not all of the dangers confronting us can be reduced through downsizing our targets. Some natural hazards we just have to face: we are unlikely to stop settling in earthquake zones, nor can we avoid tsunamis, volcanoes, asteroids, or even tornadoes. Even small targets are at risk in the case of epidemics, and terrorists with biological and radiological weapons can cause such widespread devastation that the size of the target is irrelevant.

But, except for tornadoes, all these are rare. Devastations from the more common sources such as high winds, water and fire dam-

---

from 1993 to 2001 it was 12.8, a 60 percent rise. (Special tabulation provided by Swiss Re.) Natural disasters rose steadily in this period, well below the man-made ones in the 1970s but rising to almost thirty a year for the period 1993 to 2001. Data on terrorist attacks and casualties are harder to come by, but following the end of Algerian, Italian, and Irish terrorist activity in the 1970s and early 1980s, there was a decline. But there has been a rise in the 1990s to the present. The number of "significant" international terrorists attacks has increased eightfold in the last over the last two decades, according to an analysis of U.S. government data presented in the *Human Security Report*. There has also been a clear but uneven upward trend in the number of casualties from international terror, between 1982 and 2003. (Centre 2005, 32)

age, industrial and technological accidents, and terrorist attacks on large targets can be greatly reduced. It will not be easy, but given our yearly disaster bill in the billions of dollars, it makes economic sense as well as social sense. It will require a change in our mindset about markets and regulation. Since our current mindset is only decades old—it changed quickly from the 1970s on—it is hardly inconceivable that it could not change again.

Disasters expose our social structure and culture more sharply than other important events. (Clarke 2005) They reveal starkly the failure of organizations, regulations, and the political system. But we regard disasters as exceptional events, and after a disaster we shore up our defenses and try to improve our responses while leaving the target in place. However, as Clarke persuasively argues, disasters are not exceptional but a normal part of our existence. To reduce their damage will require probing our social structure and culture to see how these promote our vulnerabilities. We will do this throughout this book in the hope of prodding changes in these areas.

Two of the major themes in this work are the inevitable failure of organizations, public and private, to protect us from disasters and the increasing concentration of targets that make the disasters more consequential. There are many explanations for the first theme, organizational failures, but we will highlight one in particular: organizations are tools that can be used for ends other than their official ones. To prevent unwarranted use, we require regulation in the private sector and representative governance in the public sectors. The failure of the political system means ineffective regulation. This can be changed.

One goal of regulation is to prevent the accumulation of economic power in private hands. Otherwise, we get the concentration not just of economic power but of hazardous materials, populations in risky areas with inadequate protection, and vulnerabilities in parts of our critical infrastructure such as the Internet, electric power, transportation, and agriculture. (We also need regulation to ensure that the public sector is not wasteful, that standards are adequate to protect us, that corruption is minimized, and so on.) The

third major theme concerns a structural alternative to the concentrations that endanger us. We encounter it first in the electric power grid and second in the Internet; these are networked systems, rather than hierarchical systems. Networks are decentralized, with minimal concentrations of destructive energy and economic power. They are efficient, reliable, and adaptive, which minimizes the dangers of organizational failures. In the grid and the Internet, they are being challenged by consolidating forces, but these can be resisted. We explore the advantages of networks in the final chapter, where we examine networks of small firms, and terrorist networks.

## THE CONVENTIONAL VIEW

In contrast to the approach taken here, which is to reduce concentrations of the things that make us most vulnerable, most efforts either accept such concentrations as inevitable or don't notice them at all and focus on responding to disasters, limiting the damages, and preventing disasters. All three strategies are necessary and should be vigorously pursued, but their limitations are apparent.

*Responding* to disasters involves "first responders," such as police, fire, and voluntary agencies. (Lee Clarke calls them "official first responders," since the friends, family, or coworkers of victims and also passerbys are always the first and most effective responders.) We have not done well here. We are barely equipped for the routine fire, flood, or explosion and fail dramatically with the big and unexpected disasters. Our new Department of Homeland Security is often criticized by the Government Accountability Office (GAO) and public policy organizations for its woefully inadequate first-responder funding. The title of a 2003 Council on Foreign Relations task force report summed up the problems: "Emergency Responders: Drastically Underfunded, Dangerously Unprepared." (Rudman, Clarke and Metzl 2003) This was apparent in the 2005 Katrina hurricane.

Furthermore, we have a "panic" model of behavior, which mistakenly limits information, which in turn breeds skepticism on the

part of the public. Years of research on disasters indicates that panic is rare, as is rioting and looting, and that the very first responders are citizens whose capacity for innovative response is impressive. The panic model unfortunately legitimates the tendency to centralize responses thus both curtailing citizen responses and interfering with the responses of the lowest-level agencies, such as police, fire, medical teams, and voluntary agencies. Research shows the most effective response comes when such decentralized units are free to act on the basis of firsthand information and familiarity with the setting. (Clarke 2002; Quarantelli 1954; Quarantelli 2001; Tierney 2003) The panic model can even make things worse. Disaster expert Kathleen Tierney shows how the media's penchant for the panic model probably caused unnecessary deaths in the Katrina disaster. (Tierney, Bevc, and Kuligowski 2006)

*Limiting damage* involves building codes, which cover structural standards and require protection of hazardous materials, and evacuation plans. There have been improvements here, but it is unlikely they will ever be enough. Government organizations often do not enforce codes; evacuation plans are unrealistic or unimplemented (as Katrina showed); inventories of hazardous materials (commonly called *hazmats*) are not made; local interests defy national standards on building sites (if these even exist); and our Supreme Court has eviscerated floodplain and wetlands regulations that would limit flood damages.

Finally, *preventing* disasters that involve vulnerable targets is the most developed of the three strategies, perhaps because there are more profits for private industry to be found in expensive prevention devices than in training and funding first responders or in following building codes. Here we have alarms and warning systems for natural and industrial dangers; and for terrorism we have biochemical snifters and suits, border and airport inspections, covert surveillance of citizens, and encryption for our electronic transmissions and interception of those of other nations. The economic opportunities here are substantial, so substantial that one angry review of the Department of Homeland Security (DHS) after the Katrina disasters spoke not of mismanagement, or even graft, but of

outright "looting" of the public treasury (Klinenberg and Frank 2005). A front-page article in the *New York Times* about Katrina and the DHS was titled "'Breathtaking' Waste and Fraud in Hurricane Aid." (Lipton 2006b) One lawyer for a prominent Washington DC law firm was up front about corporate interests: he wrote a newsletter article titled "Opportunity and Risk: Securing Your Piece of the Homeland Security Pie." (Shenon 2003a; Shenon 2003b) It is a very large pie indeed.

The Department of Homeland Security is virtually a textbook example of organizational failure that impacts all three of our disaster sources. For example, the Federal Emergency Management Agency (FEMA), once a model organization, was moved into the DHS, its budget for natural-disaster management was cut, its authority to coordinate emergency responses was taken from it, and it was staffed with inexperienced political appointees. (Staff 2005c) It became an extreme case of "permanently failing organizations" (Meyer and Zucker 1989)—those that we cannot do without but, in addition to the human fallibility of their members, are beset by underfunding in the public sphere, used for ends they are not designed for, and shackled with bad rules and regulations.

## OUR VULNERABILITIES

There is little consideration by policy makers of the possibility of reducing our vulnerabilities, rather than just prevention, remediation, and damage limitation. (Steven Flynn's book, *America the Vulnerable* [2004] is one of the few attempts to explore this. Allenby and Fink [2005] imply target reduction in their discussion of resiliency in societies.) Yet, it would be the most effective in the long run.

The sources of our vulnerabilities are threefold:

- The first are *concentrations of energy,* such as explosive and toxic substances (largely at industrial storage and process industries), highly flammable substances (e.g., dry or diseased woods, brush), and dams (one of the concentrations we can do little about).

- The second are *concentrations of populations* (in risky, even if desirable, areas), and especially when high-density populations also contain unnecessarily high concentrations of explosive and toxic substances, such as ruptured oil storage tanks in the case of Katrina and propane tank farms in St. Louis that were nearly set off by a huge flood.
- The third are *concentrations of economic and political power,* as with concentrations in the electric power industry, in the Internet (e.g., the "monoculture" Microsoft has created with the Windows operating system), and in food production such as beef and milk.

The three sources are interrelated. Concentrations of economic and political power allow the concentrations of energy, generally by means of deregulation, and these tend to be where there are concentrations of populations. To give a minor example, deregulation of the airlines started with President Carter and led initially to more airlines, more competition, and lower fares—all favorable results. But starting in the 1990s, the number of airlines decreased as the industry reconcentrated because of weakened antitrust concerns in the government. The airlines concentrated their routes through the hub-and-spoke system, which encouraged the use of ever larger aircraft as smaller flights were channeled into one hub, and passengers were then flown en masse to key destinations. Because of the concentrated hub-and-spoke system, simple industrial accidents, computer breakdowns, or false alarms at major airports now can paralyze whole sections of the nation. The attempted terrorist attack on the Los Angeles airport in December 1999 would have disrupted far more traffic than one under the regulated system. And when nature, software failures, or terrorists bring down the new airplanes carrying 550 passengers, we will realize our vulnerability to the inevitable failures of airplanes, pilots, and controllers has increased just a bit.

But the concentrated airline industry is only a relatively small instance of the increased size of our targets; we will examine larger ones in the following chapters, starting with the first of our trio of mounting threats, natural forces. The forces have not increased

much (*how* much is in dispute), but our vulnerable concentrations have greatly swollen. I will argue in the final chapter that our emphasis on the terrorist threat is exaggerated. It is a threat, but not as certain or consequential as those of natural forces and industrial and technological accidents. Were the terrorist threat to disappear, I would still recommend all the deconcentrations presented in this volume.

## THE ARGUMENT

I will start with the easiest argument in chapter 2: natural disasters, with emphasis on floods and hurricanes. Thanks to their increasing prosperity in the past half-century and a fortunate lull in patterns of extreme weather, people have moved closer to the waterfront. Their concentrated settlements are imperfectly regulated and thus are vulnerable to storms while lacking evacuation provisions. Perverse incentives in the form of subsidized insurance and federal reconstruction funds, combined with powerful local growth policies, increase our losses from natural disasters every decade. We will consider the Gulf Coast, a Mississippi River flood, and the numerous vulnerabilities of California.

Part 2 concerns the formal governmental response to disasters. Initially these were mainly natural disasters, and chapter 3 examines the government's response to disasters past. Surprisingly, the substantial role the federal government plays in helping with disasters is only about sixty years old, and I will examine that short history and in particular the role of FEMA, the key disaster agency. We will see that for a time it was an effective agency, one of the few effective ones that we will examine, but that twice in its history it has been hijacked for other purposes by presidents Ronald Reagan and George W. Bush. A model agency in the mid-1990s, it was made dysfunctional in 2001. We shall see why in chapters 3 and 4.

Disasters have other causes besides nature, and chapter 4 considers our response to terrorist threats (and the implications for FEMA). While the nation has always had domestic terrorist threats,

the scope of the 9/11 attacks prodded the federal government to do something more concerted than leaving the threat of foreign terrorist to be handled by intelligence agencies, border patrols, the FBI, and local police. This attempt by our government gives us the short and sorry history of the Department of Homeland Security, the largest federal reorganization ever. Its history illustrates well the difficulty of creating effective organizations.

We should not expect too much of organizations, but the DHS is extreme in its dysfunctions. As with all organizations, the DHS has been used by its masters and outsiders for purposes that are beyond its mandate, and the usage of the DHS has been inordinate. One major user of the DHS is Congress. While Congress is the arm of the government that is closest to the people, it is also the one that is most influenced by corporations and local interest groups that do not have the interests of the larger community in mind. Chapter 4, on the DHS and the intelligence reorganization, is mainly focused on the government, but our vulnerabilities are increasing in the private sector, and that is what we turn to in the four chapters of part 3.

Much of our critical infrastructure is in the hands of large corporations and, like our government, these private organizations are prone to error, in the form of industrial accidents as well as their failure to provide ample protection from natural and terrorist disasters. These risks are national in scope, rather than confined to an area impacted by hurricanes or floods. The private sector contains some of the largest vulnerable concentrations with catastrophic potential.

We will start with the nuclear power industry, in chapter 5, which carries the fearsome potential of a nuclear meltdown that could, as one famous report said just before the Three Mile Island near miss, irradiate an area half the size of Pennsylvania. Organizational and regulatory failures abound in this poorly regulated enterprise. Beyond the inevitable but prosaic failures of organizations, however, we will encounter here very good evidence of what will be called "executive failure," where the chief executive ignores the warnings of his staff and forces them to conduct unsafe prac-

tices. Given the size of the potential disaster in this case, such executive failures are frightening. There is a sinister side to organizational failures—not all are just prosaic—and where national safety is concerned, we should be especially concerned. As Diane Vaughan, the chronicler of space-shuttle failures, points out, this "dark side" of organizations is missing from the conventional literature. (Vaughan 1999) We shall also see how the regulatory agency bows to industry pressure and cuts back on inspections. And, of course, there is the inept attempt to deal with the terrorist threat.

Next we will examine the chemical industry in chapter 6, a vital industry that has provided better living through chemistry even while we suffer from its mostly hidden depredations. Despite its claims, it has done little to reduce the threat of terrorism, and its targets are perhaps the most poorly defended and the easiest to attack with garden-variety weapons. The danger is not just in the lack of protection from terrorists but in the huge concentrations of explosive and toxic materials, since these are also vulnerable to natural disasters, as in the case of oil spills because of Katrina. The concentrations are also vulnerable to industrial accidents, where the steady death toll could rise substantially as concentration of hazardous materials increases, along with the pressures to increase the use of unqualified contract workers in risky "turnaround" operations.

The chemical industry's record on self-regulation is not good. By claiming that they meet their trade association standards of low pollution, chemical companies escape federal inspection and pollute more. Bigger is not safer in this industry; the larger companies and the larger plants pollute more and have more accidents. Congress has been unable to set higher standards; useful bills are locked up in Republican-controlled committees. The Supreme Court declared that New Jersey's higher standards are invalid; the lower federal standards will prevail. Pressure from public interest groups has made the industry more responsive to some product safety and pollution concerns, so there is encouragement here, but these groups are not pursuing the more consequential issues of safer processes and smaller hazmat inventories.

The third conventional industry, and the single most vital one in our critical infrastructure—electrical utilities—is considered in chapter 7. The individual plants that produce electricity are vulnerable to our three sources of disaster but only moderately so. We will be much more concerned with the electric power grid. Concentration in the industry, as a result of deregulation that speeded up in the late 1990s, has resulted in business plans by the utilities that minimize investment in transmission but instead perversely overload the burdened, unmodernized grid with long-distance transmissions that reap small benefits in cost in return for large risks of cascading failures and a large increase in vulnerability to deliberate attacks, accidents, and weather. Industry self-regulation again encourages cheating on safety and reliability.

Another aspect of the grid that we will examine closely is the effects of deregulation. Before extensive deregulation, which did not lower the price of electricity but increased industry profits, the grid was remarkably efficient and reliable, self-adjusting to a limited degree, and able to handle increases in transmission without increases in blackouts. This illustrates the potential for network forms of organization rather than our traditional hierarchical forms, which will also be explored in chapters 8 and 9. These forms offer more reliability and resiliency, and less vulnerability, and could be models for other aspects of our critical infrastructure.

The public is aware of the scale and degree of the natural disaster hazard and the failures of FEMA and even the DHS, but it is largely unaware of the hazardousness of the chemical, nuclear, and electric power industries. It is even less aware of the hazard potential of the Internet, examined in chapter 8, our final industry chapter. As with the power grid, there is a resilient aspect to the Internet and World Wide Web but also a vulnerability that could make accidents or deliberate attacks very consequential. The vulnerability lies largely in the monopoly power of the Microsoft Windows operating system and the threatened consolidation of internet service providers (ISPs). Even more than the power grid, the Internet is a remarkable, vast, decentralized, resilient system that could be a model for some of our other critical industries. The threats to it

come from lack of regulation as a common carrier—we will examine this "open access" or "net neutrality" issue—and the profit potentials that come with concentration. These open the Net to easier attack, not just from viruses and hackers but from terrorists. Terrorists could disable power plants, including nuclear ones, tamper with the files of intelligence agencies, read the plans of the Department of Defense, and assault the banking sector. Hackers already have done these things.

Part 4 contains the final chapter. It brings together several of the themes and explores the costs and opportunities of deconcentrating our most vulnerable targets. One crucial question is whether the attempt to reduce the size of targets entails a loss of efficiency that is presumably gained with the economies of scale that come with large corporations. An obvious answer is that, given the disaster potential, we should be prepared to pay a bit more to reduce our vulnerability. The costs of Katrina, the 2003 northeastern power blackout, the Baltimore railroad tunnel fire, and many other such disasters greatly outweighed the expenditures that could have prevented them. A more subtle answer is to distinguish dependencies from interdependencies in our highly interconnected world. Many dependencies can and should be avoided, and instead we should reorganize to maximize the values of interdependency. In some notable cases, this increases economic efficiency as well as reducing vulnerabilities.

Chapter 9 will also revisit the terrorist threat. In the industrial chapters I have placed it more or less on a par with threats from nature and from industrial accidents. But in chapter 9 I will explore the idea that we have spent too much on the very remote possibility of another massive terrorist attack, and with little benefit given our basic vulnerabilities; and far too little on natural and industrial disasters, which are certainties; and next to nothing on target reduction. I examine the reasons why we are much more likely to experience small-scale foreign terrorist attacks than the massive ones involving weapons of mass destruction we spend most on preventing, and I argue that neither of these can receive anything like adequate protection. What will matter most is deconcentrating vulnerable targets.

Finally there is the question of whether it is even possible to deconcentrate populations, hazardous materials, and powerful organizations such as Microsoft and the electric power industry. Throughout the book I will suggest, in connection with each risky system, that this is quite possible, and I bring the arguments together at the end. Political changes over the last three or four decades have made it more difficult, but these are reversible.[2]

[2]We will not explore the ways in which the poor and minorities bear most of the brunt of disasters. They are indeed the population most "targeted" in natural and industrial disasters. The classic work on this is Bullard 1990. See also Mohai and Bryant 1992; Schnaiberg 1983; and for prescient work dealing with oil and chemical plants near New Orleans see Allen 2003. A classic broader view is Erikson 1994. For a quite different view, see "Richer is sicker versus richer is safer," a chapter in Wildavksy 1988, 59–75.

# 2    "Natural" Disasters?

------------------------------

SOCIETIES PUT THEIR people in harm's way. Modern societies do so with an especial vengeance because their technology and resources encourage risk. But sixteenth-century Holland and Venice swelled on unstable mud and have been pumping, filling, sinking piles, and building dikes ever since. Who is to say that their centuries of affluence were not worth the risk of being in harm's way? However, harm's way has been widening in recent decades. Global warming is playing a role, but the more immediate problem is development. Urbanization concentrates the targets of both "nature's wrath" and industrialization. With population growth and migration to risky areas, natural disasters worldwide are rising rapidly. More striking, the economic losses from natural disasters have risen three times as fast as the number of disasters, worldwide. In the United States, between 1950 and 1959, there were twenty major disasters costing $38 billion in 1998 dollars. But between 1990 and 1999, there were 82 major disasters, costing $535 billion. While the number of disasters multiplied by four, the costs multiplied by fourteen. The concentration of vulnerabilities has increased in the United States.

Worldwide, the total number of people affected by disasters—through loss of homes, crops, animals, livelihood, or health—almost

doubled between 1990 and 1999, affecting in some serious way an alarming 188 million people *every year*. Even wars do not take such a high toll. The figure of 188 million "is six times more than the average of 31 million people affected annually by conflict." (ISDR 2002, 49) Deaths, however, have been declining, with the advent of better warning systems and improved disaster relief services and trauma facilities. Deaths from natural disasters remains a developing-country phenomenon, a staggering 90 percent of the 880,000 estimated deaths in the 1990s occurred in developing countries—139,000 in one event alone, a cyclone in Bangladesh in 1991. There were more than 100,000 in the 2005 Kashmir earthquake. But even those do not match the estimate of 400,000 deaths in the 2004 Far East tsunami.

The most persistent devastation has been from meteorological disasters associated with water: floods, hurricanes (called typhoons or cyclones in other parts of the world), rain, and wind. The association with water is understandable. Great rivers and coasts have the most temperate climates, the fisheries, and the transportation. It is futile to intone, as one scientist does, "do not keep the water from the people, but the people from the water." (Tollan 2002) We keep the water away with human constructions, such as dams and dikes and levees, and when these fail there is disaster. But it is unrealistic to expect to keep the people from the water; salt or fresh, it is lifeblood for most of the peoples of the earth. But we can limit the concentrations and thus defend them better. Building is prevented in some areas of obvious risk; we can increase those restrictions.

Two of our major disaster areas in the United States are associated with water. One is coastal—the delta of the Mississippi River in Louisiana. The other is mostly inland—the great Central Valley Project of California, the nutrient of the state's major industry, agriculture, and the state's largest human settlement, Los Angeles. Shortly we will explore these two to examine the connection between human decisions and natural disasters and explore what might be done, but probably won't be done, about the concentrations that underlie their peril.

The natural hazards we face are many and extensive. Earthquakes and volcanoes are ever-present threats, and though a few earthquakes have been provoked by human activity (filling dams, or discharging military waste deep into the earth) (Perrow 1999), these earthquakes have been minor compared with the shifting of tectonic plates. We are unlikely to evacuate much of California, but we could disperse the concentrations of hazardous materials such as propane storage tanks and toxic chemical stocks in highly populated areas. We could also insist on better building codes and do a better job of enforcing them. San Francisco Bay is still being filled in, with large buildings resting on unstable mud that will envelope them in an earthquake, just as happened in its 1906 earthquake. Settlements on the slopes of volcanoes may be limited, but when Mount Rainer erupts as it is expected to, it is likely to devastate Seattle, despite its distance from the city. In developing countries, moreover, the poor are forced to live on the slopes of active volcanoes because the fertile land below them is reserved for agriculture by the wealthy.

Gigantic forest and brush fires have been very damaging in recent years in the United States, and these are clearly related to human activity: the suppression of natural burns that would reduce the accumulation of fuels, and allowing settlements in fire prone areas, creating new population concentrations that are vulnerable.

We will not deal with epidemics in this book because they spread so rapidly that even deconcentrated populations are at risk. But the source of some our most fearsome epidemic threats are clearly related to economic concentrations. For example, as cattle raising became industrialized, agricultural corporations herded the animals together for intensive feeding, rather than letting them graze. They raised the level of proteins in the cattle by feeding them the waste products of sheep and other animals, making mad cow disease a threat. The concentration of cattle in massive feeding pens without sanitary drainage also encouraged foot-and-mouth disease to spread easily. (Kosal and Anderson 2004) Huge poultry farms in Asia suffer from similar livestock concentrations and poor sani-

tary conditions and can contribute to the generation and the spread of such diseases as avian flu. (Lean 2006)

Damaging floods are perversely encouraged by our attempts to control rivers to protect settlements. Landslides in the United States continue to be hazards because we build on unstable slopes, and heavy rains increase the risks. In general, such vulnerabilities can be reduced in the United States through regulating structures and land-use restrictions. But there are strong pressures not to do so, and they are not easily dismissed, as we shall see.

## THE MISSISSIPPI RIVER FLOOD OF 1993

The American poet T. S. Eliot lived much of his life in England, and probably had the Thames in mind in his poem "The Dry Salvages," where he wrote of that "strong brown god" the river, but it applies to the Mississippi River as well.

> Keeping his seasons and rages, destroyer, reminder
> Of what men choose to forget. Unhonoured, unpropitiated
> By worshipers of the machine, but waiting, watching and
>     waiting.

Worshipers of the machine have cause to honor and propitiate the mighty Mississippi. It drains most of the United States—all or parts of thirty-two states and two Canadian provinces, or 41 percent of the contiguous land area in the United States, and nearly 32 percent of U.S. farm acreage. Since the 1940s, 80 percent of the wetlands, which absorb a good part of the rainfall and release it slowly into the rivers and streams that feed the Mississippi, have been converted to settlements and farmland. (Josephson 1994) The rain runs unimpeded off the streets and buildings of the settlements without being absorbed by the land, and is channeled into creeks and rivers. Modern farming leaves the residue of crops on the ground, restricting the absorption of the rain. Though the rains were just as heavy in the late 1800s and early 1900s as they were in the 1970s and 1980s, the rain has been falling more heavily in

recent decades, either as a part of a long natural cycle, or because of increased "anthropogenic forcing" (global warming from our pollution). (Kunkel 2003) Heavy spring rains in a good part of this huge area will cause occasional flooding of the Mississippi, the Missouri, the Ohio, and all their tributaries. The Mississippi flood of 1927 was worse than the one we will consider, and something of that magnitude will happen again. The flood of 1993 was bad enough, and worth reviewing since it shows how little was learned from previous floods.

The winter of 1992–93 was particularly cold, with less evaporation, so the ground was saturated. Not only was the ground saturated but a stalled high pressure system meant that the moist air from the Gulf of Mexico dropped its moisture as rain when it hit a cold front from the north. Some flooding had started in May, but the high pressure system lasted for nearly two months, in June and July. Large parts of North Dakota, Kansas, and Iowa received more than double their typical rainfall. This was the Great Mississippi Flood of 1993. Through September, nine states that fed the Mississippi had record flooding: North Dakota, South Dakota, Nebraska, Kansas, Minnesota, Iowa, Missouri, Wisconsin, and Illinois. About fifty people died and damages approached $15–$18 billion. (Hurricane Katrina's damages were estimated at $125 billion at the end of 2005.)

Nearly 150 major rivers and tributaries were affected, more than 70,000 people were evacuated, at least 10,000 homes totally destroyed, and 75 towns were completely under floodwaters, some for days. Barge traffic, carrying one-fifth of the nation's coal, one-third of its petroleum, and one-half of its exported grain, was stopped for nearly two months; ten commercial airports were flooded; and all road traffic in the Midwest was halted. In Des Moines, Iowa, nearly a quarter of a million people were without safe drinking water for nineteen hot summer days. The river widens below St. Louis, fortunately, so the run to New Orleans did little damage. (Josephson 1994; Pinter 2005; Larson 1993; Changnon 1998; Larson 1996)

An important aspect of our focus on concentrations of haz-

ardous material in risky, populated areas is that more than fifty propane tanks containing more than a million gallons of gas in South St. Louis presented the threat of a massive explosion that could have killed thousands. Propane tanks rarely explode, and the floodwaters would cool them, but the risk was not negligible. Fortunately, a levee break south of the city allowed the river level to drop around St. Louis and reduce pressure on the propane tanks. (McConnell n.d.)

We had a vulnerability of $15–$18 billion with the 1993 flood, and current weather trends that suggests it could happen again in the next few decades. Given this, it is depressing that extensive rebuilding is going forth that generally only meets the hundred-year-flood standard at best (some rebuilding is at the 500-year level) and at worst is far less, since lax local ordinances govern private and municipal levee standards. (A hundred-year record from, say, 1860 to 1960, approximately when the standards were set, picks a relatively quiescent period. Our warming climate will bring more precipitation, so that what was once a hundred-year flood may soon be a fifty-year flood.) Virtually every expert agrees that the best strategy is to limit or reduce infrastructure on floodplains. One of FEMA's singular accomplishments, in a program that President George W. Bush discontinued, was to acquire 7,700 properties at a cost of $56.3 million, including the relocation of a whole town, and turn them into water-absorbing wetlands. As geologist Nicholas Pinter sadly observes, "these buyouts are now being massively counterbalanced by new construction on the floodplains." (Pinter 2005)

Writing in 2005, Pinter says that since the 1993 flood, 28,000 new homes were built (only 10,000 had been destroyed), the population increased 23 percent, and 6,630 acres of commercial and industrial development were added on land that was inundated by the flood. In all, $2.2 billion in new development has occurred in the St. Louis area alone on land that had been underwater. To protect all this, new levees and enlarged levees are being built and the floodplain land is being raised. There is a long history of research that documents the adverse effects of levees, since they contribute

to increased flood levels. There is evidence that river channeling has increased the levels to which floods can reach by as much as eighteen to twenty-four feet during the past century in the St. Louis area. Yet they continue to be built. The floodplain laws for Missouri are among the weakest in the United States, and Missouri has passed legislation that *prohibits any county in the state from setting higher ones*. This has encouraged the extensive development in floodplain areas, which was its purpose.

Lake Okeechobee in Florida also illustrates the role of short-sighted development. The second-largest natural lake in the United States feeds the Everglades, but its water has been diverted to agricultural areas. These areas send harmful chemicals into the lake and the Everglades, destroying the ecology by encouraging algal blooms, exotic vegetation, and sawgrass. Flood-prevention actions depleted the nutrient-rich sediment that supported benign vegetation and aquatic life. The Army Corps of Engineers built a high dike around the lake after a hurricane in 1928 drowned more than 2,500 people, but now, as a result of raising the lake's level and poor maintenance, it leaks in many places and has a 50 percent chance of a major failure in the next four years. In addition to loss of lives and tens of billions of dollars in damages, it could irreversibly damage the Everglades and contaminate South Florida's water supply. (King 2006)

Some areas of the United States have imposed strict standards, but like Missouri, California has not. It has extensive risks in the Sacramento, San Francisco Bay, and Los Angeles areas. Sacramento has, according to the governor in 2005, "woefully inadequate flood prevention," and he requested $90 million in federal assistance (the Senate approved less than half of that). The aging levee system of more than a thousand miles of levees in the delta of the Sacramento and San Joaquin Rivers covers 738,000 acres of land that is mostly below sea level. The delta handles drinking water for about twenty-two million people, protects a million acres of farmland, and even housing developments on distant San Francisco Bay. Without improvements there is a two in three chance in the next fifty years that a major earthquake or even a major storm will

cause widespread failures of levees, according to scientists. (Murphy 2005) Similar development has occurred around Dallas, Kansas City, Los Angeles, and Omaha, says Gerald Galloway, a professor of engineering at the University of Maryland. (Bridges 2006)

Geologist Jeffrey Mount notes an irony: the farmer who makes a modest investment in shoring up the levee protecting his farmland will see the property value rise, increasing the likelihood that the farmer will sell to developers, putting more people and more property at risk when the levee breaks. (There are two kinds of levees, another expert notes: those that will fail and those that have failed.) (Bridges 2006)

Congress, the Bush administration, and the U.S. Supreme Court have all weakened wetlands preservation since 2001. In January of that year the Supreme Court ruled that wetlands not directly connected to other water bodies are not protected by the Clean Water Act regulations that restrict their development. Cooperation between states linked by rivers has been difficult. In marked contrast, Europe has much more far-reaching projects and standards. The Dutch government, which had a 1953 flood comparable to the one caused by Katrina, has a policy of "more room for the rivers," which means creating new storage and conveyance space rather than a new round of levee-raising. Villages have been relocated. Cooperation between nations in Europe has been better than cooperation between states in the United States. There, France, Germany, Belgium, Luxembourg, and the Netherlands—all affected by the Meuse River—have a plan for longer storage and slower release and more space for the river.

The United States is making some progress. Pinter observes: "Thanks to federal guidelines, buyouts, and enlightened management in many localities, successes in managing U.S. floodplains outnumber the failures. The problem is that when these measures succumb to local economic self-interest and political pressure, small local failures—like cracks in levees themselves—allow massive increases in floodplain infrastructure that can rob the nation of all of the net improvements painstakingly won elsewhere." When the floodplains disappear behind new and larger levees that

allow development, the more farsighted communities downstream will be overwhelmed. Citizens of Arizona and Colorado, who pay for better flood management, will pay again when the federal government bails out Missouri, where legislatures choose development over safety.

## KATRINA AND NEW ORLEANS

New Orleans is built on ground that is slowly sinking into the Mississippi delta. Much of the city is inside a ring of concrete and earthen sides, rather like a bowl, fourteen feet below the surrounding waters at its deepest, and the bowl is attached to a long string of highways wandering through the sinking Mississippi delta. The bowl is surrounded by water, and the water is higher than many parts of the city that sit on the floor of the bowl. Adjacent to the city is a very large basin, Lake Pontchartrain, which is itself only a foot above sea level. Tourists climb the many steps up from a street to look at the top of the lake. Cutting through the town is the Mississippi River, itself higher by two to five feet than the historic French Quarter. Tourists regularly wonder at the superstructure of giant ships moving slowly above them as they strolled the streets.

The Louisiana delta is the city's protection, and it is disappearing, no longer fed by the farmlands of nine states. The dams and levees concentrate the force of the river's flow, a form of economic concentration we ordinarily do not consider. Channels below New Orleans carry the silt far out into the Gulf of Mexico in a concentrated stream. The rest of the delta, starved for replacement soil, subsides. "Since the 1930s, an area the size of Rhode Island has sunk beneath to waters of the Gulf of Mexico." (Scully 2002) Every hour a parcel of wetlands the size of two football fields converts to open water, with incursions from the Gulf. Every 2.7 miles of marshland that a hurricane has to travel over reduces the surge tide by a foot, dispersing the storm's power. Simply put, had Katrina struck in 1945 instead of 2005, the surge that reached New

Orleans would have been as much as five to ten feet shallower than it was. (Tidwell 2005)

What was lost no longer provides a break to the wind in storms, or to the surge of seawater the wind produces. So New Orleans gets more wind and more water, and builds higher levees (and taller high-rise buildings).

A report from the Intergovernmental Panel on Climate Change (IPCC) notes that extreme droughts can cause saltwater "wedges" to move upriver (such as occurred in 1988), threatening drinking water supplies. Even heavy rain events are consequential; they can increase the flow volume in the Mississippi and Atchafalaya Rivers, threatening navigation and flood-protection structures. (IPCC 2002, 24) When we channel the silt-laden water of the brown god to the Gulf to prevent seasonal flooding and build ports and cities on its porous shores, we find the cities sinking as they are starved of the silt. Such is the case with New Orleans; it steadily sinks.

All this has not gone unnoticed but has been carefully studied and publicized for at least four decades. *National Geographic* ran a handsome, scary spread in 2004, "Gone with the Water." (Bourne 2004) In 2002, a special two-part series in the *New Orleans Times-Picayune* foresaw much of the destruction of Hurricane Katrina. (McQuaid and Schleifstein 2002) It noted that if a storm surge topped the lake's levee system along its south shore, the death toll could be between 25,000 and 100,000, leaving thirty feet of standing water (two and a half stories). Katrina's damage, as awful as it was, was paltry compared to what could have happened. It breached levees and leaked water that slowly rose, rather than overtopping them in a sudden surge that would leave no time for people trapped in their houses to go up to the attic.

As it was, the *Times-Picayune* story foresaw many aspects of the destruction Katrina caused. One particularly interests us, because it concerns the concentrations not just of populations but of hazmats. It quotes an expert: "We don't know where the pipelines are, and you have the landfills, oil and gas facilities, abandoned brine pits, hardware stores, gas stations, the chemicals in our houses," said Ivor van Heerden, assistant director of the Louisiana State

University Hurricane Center. "We have no idea what people will be exposed to. You're looking at the proverbial witch's brew of chemicals."

Southern Louisiana is a major shipping center and industrial area (for mostly toxic and explosive industries such as chemicals, oil and gas). Southern Louisiana produces one-quarter to one-third of the country's seafood, one-fifth of its oil, and one-quarter of its natural gas, so the potential economic loss from a more serious blow will be enormous. The chemical and energy companies, of course, has known all along of how their steady expansion endangers their economic well-being, but expansion drives market capitalism. The obvious precautions they take are for their individual installations, not to protect the wetlands that would reduce overall risk. Most of the region's chemical plants and other large industrial facilities are relatively safe from flooding, the *Times-Picayune* account asserts, without convincing evidence. The evidence is now coming in that the destruction of pipelines and other infrastructure from Katrina was far more severe than expected. It will be some time before we know the full extent of this kind of damage from Katrina, but we do know about the oil spills, and the case of Murphy Oil is sobering.

### Katrina, Oil Spills, and Murphy Oil

Hurricane Katrina demonstrated quite clearly the danger of locating hazardous materials near urban populations. Katrina's strong winds and crashing waves dislocated oil storage tanks at numerous New Orleans oil facilities from their foundations, causing millions of gallons of crude oil and other chemicals to leak into the surrounding areas. The Natural Resources Defense Council tracked 575 petroleum and hazardous chemicals spills. Ten of the biggest spills in Louisiana add up to about eight million gallons; by comparison, the *Exxon Valdez* accident released about eleven million gallons into a remote area in 1989. (McKay 2005) In the case of the Murphy Oil Corporation headquartered in El Dorado, Arkan-

sas, the surrounding area was the St. Bernard Parish of New Orleans, a town of 70,000 residents. Roughly 1.1 million gallons of
oil spilled from Murphy's tanks into St. Bernard Parish, coating
1,800 homes with a thick layer of black crude oil, making them
worthless, and causing high levels of property damage in a wider
area. Residents of St. Bernard returning to their homes were exposed to benzene and other harmful chemicals contained in crude
oil, and many suffered short-term health effects including nausea,
dizziness, and skin irritation.

After the spill, Murphy Oil hired a private contractor to begin
cleaning up the spilled oil. However, because the law requires that
Murphy obtain permission from individuals before cleaning their
property, many homes went untouched for weeks because their
owners were dispersed and could not be contacted, allowing crude
oil to soak into the residents' yards. Oil that is saturated into buildings or soil is very difficult to clean, and, according to the EPA,
could be carcinogenic in the long run. Because of this spill, the
1,800 households covered by Murphy's oil face daunting cleanup
tasks and health risks in both the near and distant future. (McKay
2005)

Did Murphy properly prepare for Katrina's onslaught? A common, but not legally mandated, procedure for oil companies prior
to hurricanes is to fill their tanks with either more crude oil or
water, thus weighing them down so that flood waters do not rip
them from their foundations. Murphy, however, filled its 250,000-
barrel storage tank with only 65,000 barrels of crude before Katrina struck. Predictably, the tank was dislodged by floodwaters
and over 25,000 barrels of crude leaked out of the tank. Murphy
has been vague about its internal hurricane guidelines and whether
they were met before Katrina, and has simply blamed the spill on
"an act of god." (Burdeau 2005) While no preparation can guarantee resistance to a hurricane, Murphy's proximity to a residential area, combined with relatively empty storage tanks, certainly
contributed not only to a massive loss of oil, but to a nightmare for
homeowners who found their homes awash in filthy black crude.

Although Murphy's failure to fill its tanks to capacity reduced

the chance that they would survive Hurricane Katrina, even vigilant application of this tactic cannot definitively protect against massive spills. The Cox Bay facility of Bass Enterprises filled its two tanks half-full with 45,000 barrels of oil before Katrina struck. This level is 15,000 more barrels than company standards call for and a standard that had succeeded for fifty years. (Cappiello 2005) Yet the Cox Bay facility suffered an even greater spill than Murphy Oil's St. Bernard facility: 3.78 million gallons of oil poured from its tanks when Katrina dislodged them from their foundations. (U.S. Congress 2005) Clearly, the industry standard is too low.

Fortunately for Bass Enterprises, much of the oil was contained in dirt berms around the site, although one containment reservoir designed to hold spilled oil was already full of water by the time oil began to leak. (Cappiello 2005) Perhaps more important, the Cox Bay facility is in the St. Plaquemines Parish, a town of only about 20,000 residents, compared with the 70,000 of St. Bernard Parish. The tanks themselves are located right on the Mississippi River, not directly next to a residential area. Because of this positioning, Bass Enterprises was able to spill even more oil than Murphy without covering thousands of homes in oil. In fact, Bass was able to contain or recover roughly three million gallons of the spilled oil, while most of the remaining oil simply evaporated. (U.S. Congress 2005) Although lamentable, the harmful effects of this spill paled in comparison with the much smaller spill at Murphy.

### Reclaim the Delta?

Reclaiming the land would cost many billions in a state that is desperately poor with regressive taxation that favors business and industry. According to the *Houston Chronicle,* a consortium of local, state, and federal agencies in 2001 were studying a $2–$3 billion plan to divert sediment from the Mississippi River back into the delta. This would replace only 2 percent of the loss, however. Other possible projects include restoration of barrier reefs and perhaps a large gate to prevent Lake Pontchartrain from overflowing and

drowning the city. There is little state money for these multibillion-dollar projects, so it was doubtful that much would be done, and it wasn't. In 2001, according to the *Chronicle,* "a plan to restore the Florida Everglades attracted $4 billion in federal funding, but the state had to match it dollar for dollar. In Louisiana, so far, there's only been a willingness to match 15 or 25 cents." (Berger 2001)

More alarming still is the decision of President Bush to effectively dismiss a proposed $14 billion delta rescue plan; he was willing to authorize only less than 2 percent of that: $250 million. (The early estimates of the cost of Katrina are $125 billion.) The proposed plan was widely viewed as technically sound and supported by environmentalists, oil companies, and fishermen alike, and had been on the table for years and was pushed forward with greater urgency after Katrina hit. According to one source, it "would use massive pipelines and pumps and surgically designed canals to guide a portion of the river's sediment-thick water back toward the coastal buffer zone without destroying existing infrastructure or communities. This would rebuild hundreds of thousands of acres of wetlands over time and reconstruct entire barrier islands in as little as twelve months. (It is estimated that the government's plan to rebuild the levees could take decades.) Everyone agrees the plan will work. The National Academy of Sciences confirmed the soundness of the approach just last week and urged quick action." It is the equivalent of six weeks' spending in Iraq, or the cost of the "Big Dig" in Boston. (Tidwell 2005) Spending 10 percent of the estimated cost of the hurricane on reclamation would have made the city safe and the delta once again a rich food source.

### Rebuilding New Orleans

Louisiana is at or near the bottom of most indices of social welfare, social well-being, and education and ranks high in political and police corruption. Can we expect it to rise to the challenge of its mounting vulnerability to meteorological disasters without drastic downsizing? Probably not. True, the largest port in the United

States cannot function without a city around it. But we need a city of perhaps one-quarter its pre-Katrina size to support about 10,000 people in the two critical industries: the port, and oil and gas. The city has been losing population slowly since 1960, in part because of mechanization of the gas and oil industry and the port. There were fewer than 7,500 people working in the port, fewer than 2,000 working in oil and gas extraction, and fewer than 100 working on pipeline transportation before the tragedy. (Glaeser 2005) Protecting a small city from hurricanes would be easier.

But the problem is the unprotected infrastructure, estimated at $100 billion. This includes important ship channels such as the Gulf Intracoastal Waterway, major ports such as Lake Charles and New Orleans, roads, and oil and gas pipelines and facilities. (IPCC 2002, pp. 19–20) Most of this infrastructure was designed to function behind a buffer of protective wetlands, not for full exposure to the Gulf. The Intergovernmental Panel on Climate Change, in its section dealing with Southern Louisiana, was far from optimistic in its 2002 pre-Katrina report. It saw an inevitable retreat, involving the relocation of inhabitants. The costs of retreat would include lost incomes, relocation and adjustment expenses, and losses in quality of life. Along with financial costs would come social and cultural costs. It quoted one of its studies: "It is almost impossible to quantify the value of a loss of lifestyle, and loss of family and social ties which will inevitably come as wetlands disappear and prompt people to relocate." (IPCC 2002, 22) With Katrina, the relocation is under way and the social impact is devastating.

### It's Not the Weather; It's Affluence and Opportunities

Louisiana is only a particularly dramatic example of coastal erosion in the United States. Worldwide it hardly compares with the erosion and risks of the other major rivers on the earth. In the United States, as in the rest of the world, the problem is exceedingly simple: rather than keeping people away from the water,

people are trying to keep the water away from them. Water, as noted, is the prime location factor for human populations, and humans interfere with nature's regenerative cycle. With the population scale and technological power we have achieved, we invite disasters. In an article titled "Why the United States Is Becoming More Vulnerable to Natural Disasters" the authors demonstrate that "more people and more societal infrastructure have become concentrated in disaster-prone areas." (van der Vink et al. 1998) The costs of natural disasters in the United States were estimated to be $1 billion per week, or $54 billion per year, before Katrina's $125 billion, so the cost is of some concern. Mitigation efforts have reduced the loss of life, but the economic losses have been soaring since the mid-1980s. This was not because of any marked increase in the severity or frequency of natural events; hurricanes had not increased in the 1990s (though they have since then), nor have earthquakes in the last century, and the slight increase in tornado sightings was probably due to better reporting. What does correlate highly with disaster costs, however, are population changes—the movement of people to disaster-prone areas and the growth in revenue of such areas, measured as the total common tax revenue of the state, an indicator of the wealth that is put at risk. The study uses crude measures—all of California is lumped together, for example—but the correspondence of disaster costs on the one hand, and population growth and wealth growth on the other, is striking and increases over the measured period, 1970 to 1995.

"States most affected by the costs of hurricanes (Florida, Maryland, North Carolina, and Texas) and earthquakes (California and Washington) show the largest increase in both population and revenue," research has found. Why have the increased costs of natural disasters not created pressures to encourage proper land use in vulnerable areas? Because "[e]conomic incentives for responsible land use have been stifled by legislated insurance rates and federal aid programs that effectively subsidize development in hazard-prone areas." Though they do not go further than this, the authors

might have added that since, as they say, the people moving into the risky areas "represent the higher wealth segments of our society," they are likely to have the most political power, and favor subsidized development in risky, though beautiful and profitable, areas. (van der Vink et al. 1998)

Climatologists Stanley Changnon and David Changnon, funded by the Electric Power Research Institute, have more specific data. (Changnon and Changnon 1999) They examined the *insured losses* (a lower figure than total losses) of more than $100 million per incident from 1950 to 1996. The last six years of this period, 1990–96, showed average losses much higher than did the 1950–89 period. Was it weather or social processes? The more recent period did not have more intense damaging events (mostly hurricanes), or larger ones, or even greater losses per event. Only coastal storms in the East and hailstorms in the West have increased in recent years, but not enough to explain the sharp increase in losses in the 1990–96 period. Two demographic shifts have put people in harm's way. First, people have moved to the coastal areas of the South and the East; the southeastern coastal areas had a 75 percent increase in population density from just 1970 to 1990, while the national increase was only 20%. Population also increased in California and Washington, both vulnerable states. Second, people have also moved into cities. In 1960, 58 percent of the population lived in metropolitan areas, but by 1993 it was 79 percent. This has meant higher density of structures and vehicles, which produces large losses even with small-scale, but intense, storms. (Shoddy practices played a role. They quote a study revealing that improper construction practices accounted for $4 billion of the insured losses of $15 billion from Hurricane Andrew in 1992.) They conclude: "Assessment of the effects of growing population, demographic shifts, increasing property values, and poorer construction practices suggests that these 'societal factors' have collectively acted to make the U.S. ever more vulnerable to damaging storms and are the principal reasons behind the record high number of catastrophes and associated losses of the 1990's." (Changnon and Changnon 1999, 299)

### Can Anything Be Done?

What is to be done? The vulnerable areas are filling up with poisons and people because of countless personal choices, it is customary to say. But we can usefully sort out the forces behind these choices. Some of the vulnerability is due to changes in income distribution; with more of the rich able to get good views in warm climates, places such as the Florida Keys and San Diego expand. The rich who live there at the time of a disaster are at risk, of course, but the poor who service their lifestyle are at even greater risk, in fragile homes and unable to fly out. In New Orleans the poor could not even drive out; the best the city could do was to prepare a DVD warning them that they would be on their own, and that was not distributed in time. It was left to the churches to try to arrange carpools, but this effort, long planned, had not gotten off the ground before Katrina struck. (Noland 2005)

The Florida Keys are an example of perverse cross subsidization: the poorer parts of the state and the nation are taxed to support government programs to develop the area's essential infrastructure and subsidize the insurance payments. Only a political change (unlikely) could redistribute wealth and charge residents and their merchants the true cost of maintaining them there. Absent that, government would have to impose stricter standards on design and construction, which is hardly controversial and could do some good.

Industry is another matter. It is legitimately drawn to coast lines because of the oil, gas, fish, and favorable agricultural conditions. The major problem here is capitalism's dynamics, which favor short-over long-run investments. Even an oligopoly or a monopoly is restrained from taking a long-run view if it is publicly traded. Investment funds are not likely to take a long-run view by holding on to a stock that is low performing in the short run, in the expectation that this will pay off in the long run of ten or twenty years. Though ideally performance is understood in terms of long-run expectations—just look at Internet firms, which have high stock values even though they have little revenue, and conservative

retirement funds—the market does a poor job of anticipating possible disasters. It is difficult for the large investment funds to tell their investors that this low-performing asset is still valuable because it is protected from risks that just might materialize in ten to twenty years, whereas investments in hot stocks are not. The fund would not perform at its maximum. So the investment firms withdraw funds from the lower-performing companies and put them into the higher-performing companies. The higher-performing companies in any particular industry are likely to be focused on the short run. In the long run they will suffer from not investing in safety, but a few miles away there will be another short-run company that was missed by the storm, and it will see its stock rise. If investing in safety would improve its quarterly or yearly rate of return, the company will do so; if the investment (say rebuilding pipelines that are becoming hazardous because the sinking land has exposed them to wind blown water) might pay off only if a storm hits that particular area in the next twenty years, it will be reluctant to do so, especially if the executive making the decision is to retire in five years. Industry and agriculture will have to remain in vulnerable areas, but we should not rely on a short-run market mentality to police it.

## CALIFORNIA'S CENTRAL VALLEY

Compared with the disaster potential of Los Angeles, California, the actual disaster in New Orleans is small. Los Angeles is a mega city, with half the population of California within a sixty-mile radius of its politically dysfunctional center, the product of massive continental and intercontinental migration typical of globalization processes. (Mexico City understandably has the highest concentration of Mexicans in the world, but Los Angeles has the second.) And it is the product of the massive relocation of the main natural resource for humans: water. With a wretched system of governance—including a lack of intermediary levels of governance to encourage citizen participation—its impoverished millions are ha-

rassed by its economic elite. The list of economic and social problems of Greater Los Angeles (about sixteen million inhabitants)
tops that of any U.S. city. But that is not our concern here. Rather,
we want to know about its vulnerability to natural disasters and
how its process of growth has contributed to the vulnerability.
Truly significant reductions in vulnerability would require the
deconcentration of organizations and populations, and that is not
likely to happen; only minor deconcentrations may be beginning.
Our initial guide will be a chapter by Ben Wisner from the book
*Crucibles of Hazard: Megacities and Disasters in Transition* (1999).
But the classic litanies go back to Carey McWilliams (1949) and
Mike Davis, especially his *City of Quartz* (1990).

The earthquake hazard is the most obvious. Los Angeles lies on
or near two major fault lines, and "the big one" is expected within
decades. (Predictions are obviously very difficult. In 1985 a large
earthquake was predicted to occur in the Parkfield area before
1992, with a 95 percent probability, and it has yet to arrive.) Building codes and site restrictions are improving only slowly and
would make little difference if the quake is massive and breaks any
of the numerous dams aimed at the city on a desert.

Los Angeles would be a small agricultural service center if it
were not for massive state and federal investments in the twentieth
century; it has no intrinsic value other than its weather, which it
shares with much of the coast in the area. But water was brought
in from hundreds of miles away, and state and especially federal
aid have provided sewage treatment plants, a deepwater port
dredged at nearby San Pedro, the Boulder Dam for water and
power, the San Onofre nuclear power station, and the massive free-
way system.

California's Central Valley Project, without which Los Angeles
would be a smaller city, has been called the largest public works
project in the world. There are a series of more than twenty dams
and their reservoirs, five hundred miles of canals stretching much
of the length of the state, and, as noted, miles of aging levees protecting dense populations. The dams started out as a flood-control
project, but were expanded by those interested in hydroelectric

power, recreation, farming, and a water supply. The huge dams prevented flooding and irrigated the desert, which bloomed with produce and made agriculture the leading economic activity of the state. Fully 85 percent of the heavily subsidized water is used for agriculture.

The key to Los Angeles' vulnerability is the agency that determines what shall be developed. The Community Redevelopment Authority, appointed by the mayor, is largely composed of businessmen, and able to designate development areas, to commandeer land by means of compulsory purchase, and to raise money. It generates handsome revenues, only 20 percent of which are required to be spent on low-income housing; the rest goes to middle- and high-income housing and economic development projects—that is, private business. (Wisner 1999)

Economic development includes the chemical industry. The Los Angeles Standard Metropolitan Statistical Area (SMSA) has the second-highest number of chemical facilities in the United States; this is a spectacular concentration. Many of these, including the major oil refineries, are located near earthquake faults, particularly the Newport-Inglewood fault. Automated systems for shutting down risky facilities are required, but those familiar with accidents in risky systems are mostly bemused by these requirements—they are worth having, of course, but the possibility of interacting failures that can defeat safety systems is ever present. In 1987 an earthquake in the Whittier Narrows area displaced a one-ton chlorine tank that was being filled, releasing a half-ton toxic cloud. The power failure caused by the earthquake disabled the company's siren, and malfunctioning telephones made it impossible to warn authorities. (Showalter and Myers 1992) It could have been a ninety-ton tank car. There are undoubtedly other such instances, but it is surprisingly difficult to document toxic releases associated with disasters, and it is quite unclear as to how much damage this toxic cloud caused. If it had been a ninety-ton railroad car, a few million people would have been at risk.

Much has been done to upgrade building standards and prepare for emergencies, since the risks of earthquakes are obvious. After

the 1989 Loma Prieta earthquake, strict state building codes for new housing were established, and ambitious hazard abatement programs for bracing existing buildings were instituted. But relocating facilities with toxic and explosive substances, especially chemical plants, oil refineries, and natural gas facilities seems to be out of the question. Why they were built there in the first place and allowed to expand, since the hazards were known and risky areas identified many decades ago, is an important question. The answer appears to be economic pressures on the political representatives responsible for overseeing decisions about site selection, protection, and the enforcement of regulations. Market forces are allowed to control growth, but there seems to be a market *failure* for preventing or mitigating low-probability/high-consequence accidents. Santa Barbara and some communities in Northern California have enacted growth controls that reduce risk, but that has been comparatively easy in these wealthy communities where industry is absent. (When industry invaded Santa Barbara from the sea with oil spills in the Channel in the 1960s, we witnessed the spectacle of upper-class conservatives demonstrating in the same manner as the New Left hippies they despised.) (Molotch 1970)

## DECENTRALIZATION AND REGULATION

I have argued for decentralization, but it has to be accompanied with a reduction in what economists call (unfortunately) "information asymmetry," where one party knows more than the others. What we need are a representative governance mechanism that ensures full disclosure and standard-setting and coordinating mechanisms. Decentralized government without these has left local authorities—legislators, city officials, planning boards made up of prominent citizens—open to local property-development interests. These interests are able to get unpublicized waivers on building-code requirements even if, occasionally, stringent standards exist. If the planning boards and even the elected officials are unrepresentative, private interests will be served at the expense of com-

munity ones. If federal standards are weak and unenforced, private interests will prevail. Private interests appear to have prevailed in robbing New Orleans of its protective wetlands.

A dilemma that we will repeatedly encounter is that, in the interests of decentralization (e.g., giving states authority rather than the federal government), the federal government cannot set higher standards than the states. Occasionally the federal government sets standards that are lower than a state desires and these lower standards prevail; sometimes local standards may be allowed to prevail over higher state standards. Federal and state governments should establish *minimum* standards, which states or localities can exceed.

It has taken two centuries to establish effective minimal federal standards in health, human rights, financial, and electoral processes. Where they have succeeded there has been a recognition that an increasingly complex and tightly coupled society cannot survive when critical areas that invite self-interested behavior are radically decentralized. It is these areas this book is concerned with, and generally I argue for federal standards, but only if they are higher than state or local ones. But interests can legitimately argue that federal standards are inappropriate—they are too expensive, they involve unexpected impacts on legitimate interests, they maximize the wrong values, and so on. This is a political problem, and over the centuries the decisions as to what federal standards should be are more political than technical. But some proposed standards are so sensible they hardly seem to warrant contestation, and many of these will be advocated in this book. With regard to threats of natural disasters, they include higher building standards and greater restrictions on land use than the federal government has established. Where states and localities deem the federal standards to be too low, state and local authorities should be allowed to raise them for their jurisdictions, but the U.S. Supreme Court can hold otherwise, as it has more than once. This has allowed private power to resist higher standards and thus endanger buildings and settlements in risky areas.

"Few local governments are willing to reduce natural hazards by managing development," observes disaster researcher Denis Mileti,

whose work we will draw upon here. They have more pressing concerns than mitigation. "Also, the costs of mitigation are immediate while the benefits are uncertain, do not occur during the tenure of current elected officials, and are not visible (like roads or a new library). In addition, property rights lobbies are growing stronger." (Mileti 1999, 160) (Since 1999, when he wrote this, court decisions have greatly strengthened property rights, to the dismay of not only environmentalists but disaster specialists.)

It was not until 1994 that a model building-code council was established for the United States, and even so, adopting the model code is voluntary. Local and state codes are weak; only half of the states have state codes that local authorities must use, exemptions are easy to obtain, and enforcement is often lax. South Florida was thought to have one of the strongest codes in the nation, but Hurricane Andrew showed it was not properly enforced, and a 1995 survey of building department administrators in the southeastern United States found that half did not have the staff to perform inspections. (Mileti 1999, 165)

We need a more centralized regulatory system. Local initiative is simply not reliable in the case of mitigation. Localities are reluctant to enforce state standards, national standards are few, and enforcement is lax. This is an area where centralized regulation—standards and enforcement—is needed; given the political influence of growth-oriented city officials and property and building interests, there is bound to be a "failure" of the private market.

Insurance, both from government and private sources, could be the most powerful tool to ensure responsible action by individuals, organizations, and governments. But it represents another failure. Even in areas known to be hazardous, only about 20 percent of homeowners purchase flood insurance, and less than 50 percent of businesses purchase flood and earthquake insurance in risky areas. (168) Those that do can benefit handsomely from federally subsidized insurance if there is a disaster. The tiny community of Dauphin Island at the entrance to Mobile Bay, Alabama, has been hit by six hurricanes in the past twenty-five years. The community has collected more than $21 million from the nation's taxpayers.

Katrina almost demolished the barrier island but the inhabitants were not denied coverage. A study by the National Wildlife Federation pointed to a homeowner who filed sixteen claims in eighteen years and received a total of $806,000 for a building valued at $114,000 from federal flood insurance. (Staff 2006b)

Federal assistance programs are not as generous as our National Flood Insurance Program. While the assistance programs are thought to be equivalent to hazard insurance, they are very small ($3,000 average in recent disasters in the case of FEMA's Individual and Family Grant program) and intended only for immediate needs. Still, the existence of these federal assistance programs is cited as an important reason why people continue to settle in hazardous areas, making the programs perverse. (Mileti 1999, 169) It is hard to keep people from the water; federal assistance does not help.

Insurance companies are in a bind with regard to large-scale catastrophes: the number of catastrophes have increased, capital markets are more unstable, and the insurance regulatory system provides perverse incentives. Before 1988, for example, insurance companies had never experienced a loss from a hurricane that exceeded $1 billion; the risk was manageable and spread over wide areas. Since 1988, there have been sixteen disasters with losses exceeding $1 billion; Hurricane Andrew resulted in losses of $15.5 billion, even though it bypassed the most heavily developed parts of the Miami metropolitan area. (170) Losses from Katrina are estimated at $125 billion. With losses so high, and risks growing in Florida and California (following the Northridge earthquake), insurers have tried to withdraw from these policies in these states, but some are unable to do so. Perversely, some states that require them to insure do not establish or enforce standards.

State-mandated pools have been established to serve as a market of last resort for those unable to get insurance, but the premiums are low and thus these have the perverse effect of subsidizing those who choose to live in risky areas and imposing excess costs on people living elsewhere. In addition, the private insurers are liable for the net losses of these pools, on a market-share basis. The more

insurance they sell, the larger their liability for the uninsured. Naturally, they are inclined to stop writing policies where there may be catastrophic losses (hurricanes in Florida and earthquakes in California). (171) The Florida and California coastlines are very desirable places to live and their populations have grown rapidly, but these handsome lifestyles are subsidized by residents living in the less desirable inland areas in the state and, to some limited extent, by everyone in the nation. Every area has its hazards, but the fastest-growing areas have greater ones. A national policy of vulnerability reduction would stem growth in these areas by making them pay their fair share of catastrophic loss insurance, fewer people would mean fewer unavoidable deaths, the economic incentives to build more safely would increase, and the result would be fewer avoidable deaths.

## CONCLUSIONS

We are tempting nature by putting concentrations of hazardous materials, populations, and vital parts of our infrastructure in its way. (We are also increasing nature's fury with our disproportionate contribution to global warming, but that is not the topic of this book.) The strong central government of Holland protects its citizens from thousand-year floods and creates wetlands for runoffs from the yearly surges of its rivers. In the United States, development in our midwestern river valleys and our Gulf Coast is removing the Gulf Coast wetlands at the rate of two football fields every hour, bringing storms and the ocean ever closer. In contrast to Holland, private and municipal levees in Missouri do not even have to cope with hundred-year floods. The levees controlled by the state of Missouri are thinking of five-hundred-year floods at best. Even our levees in New Orleans were unable to handle a category three hurricane, and a slight shift of Katrina could have suddenly inundated the city, producing perhaps ten times the number of deaths that it had from a few slow leaks.

A recent Supreme Court ruling exempted many low-lying areas

in the nation from a designation of floodplains, allowing building to proceed. Private property-rights legislation and Supreme Court rulings have replaced the community as the referent with the individual business or landowner as the referent, so communities are less protected. Huge oil tanks are not filled to keep them stable when hurricanes are predicted, and a large farm of propane tanks barely escaped detonation in a Mississippi flood in crowded St. Louis. Not only weather can attack the concentration of hazmats and the surrounding populations; it could be an industrial accident, such as an explosion or train wreck; or it could be a team of terrorists. Building bigger levees increases the scope of the inevitable disaster, while it brings in the business and populations who are protected from minor storms but made more vulnerable to major floods and hurricanes. Our wealth pushes us up against hills that can produce mudslides or into beautiful canyons that can be swept by raging forest or brush fires.

And we subsidize these risky delights by taxing those in safer and less beautiful environments. As we shall soon see, in our politically fueled preoccupation with remote terrorist dangers, we cut the federal funds for prevention, response, and remediation from natural disasters.

All of this is avoidable; much of it can be mended. Some of our dangerous practices have been slowly accumulating over the new century, but they can be reversed with strong regulations and strong liability laws. However, some of the dangers are the result of public-policy shifts that appeared in the 1970s and 1980s. The regulations that once protected us have been weakened, and the warnings from experts about the inappropriate policies in our recent building binge have been ignored. In the next chapter we look at where government has been in all this. Borrowing a line that Senator Charles Schumer (D-NY) applied to the response to the 2003 Northeast blackout, Katrina was not even a wake-up call; we have hit the snooze alarm instead.

**PART 2**

**Can Government Help?**

------------------------

# 3 The Government Response
## The First FEMA

------------------------------------

THE FEDERAL RESPONSE to disasters has been predictably late, reactive rather than proactive, accommodating to the special interests of business and Congress, and unable to enforce penalties for failing to take obvious precautions. This is predictable for at least two reasons. One is that there is no unified constituency for the variety of disasters that can occur that would demand planning over a long time horizon; consequently, the government is often reacting to rather than anticipating disasters. As federal involvement increased in the second half of the twentieth century, each major event fostered new laws and the creation of new organizations, sometimes to be folded into the next reorganization, or just left hanging loose, trying to hold on to their constituency. This evolution does not invite proactive, long-range planning.

The second problem is that there are too many constituencies who are competing among themselves and unwilling to make the changes that reform would require. Disasters mean money will be passed out and political decisions made about rebuilding; many different players are attracted to the disaster scene. The large sums given out invite corruption, but more important, the emergency and rebuilding funds can be distributed according to political and economic interests rather than need. These interests include the

terms of rebuilding such as condemnation of buildings and of land, building codes, and zoning ordinances; new equipment for official first responders (fire, police, etc.); and insurance requirements and regulation of the insurance industry. These reflect the interests of government agencies and their constituents, and of economic elites in general. And finally, the succession of organizations set up to deal with disasters creates organizational opportunities for unrelated political ventures (e.g., limiting personal privacy, privatization of government) and economic ones (e.g., firms marketing their products).

Thus, there is little incentive, because of the short-term interests of organizations in the public and the private sector, to utilize a long-term perspective that would assess the built-in vulnerabilities of the nation, such as concentrations of hazmats or populations that cannot be evacuated, so we are reactive rather than proactive. Short-term interests seize on disasters as opportunities to be exploited. It is a basic premise of this book that organizations are tools; they can be used for many things other than their official mandate. Since disasters involve most everything in society, the interests that can be served by organizations involved in disasters are broad and powerful.

We do learn, however, and regardless of the interests and uses, the U.S. federal government has evolved institutions and organizations that cope to some degree with our infrequent but inevitable natural disasters.

## EXPANSION OF FEDERAL RESPONSE

There was no official mandate for the federal government to respond to disasters through most of our history as a nation. It is really quite new. David Moss (1999) examined the history of federal disaster relief in terms of three phases of federal protective services. The first, in the nineteenth-century, was concerned with creating a secure environment for business and involved legislation dealing with such things as limited liability, property rights, bankruptcy,

and deposit and crop insurance. The second, from 1900 to 1960, emphasized creating a secure environment for workers. This involved workers' compensation, old-age insurance, unemployment insurance, and disability insurance. A regulatory system was gradually added covering occupational safety, health, pension regulation, and insurance regulation. From 1960 on, federal involvement increased substantially, largely as a result of increased prosperity and rising standards of living. Moss calls this "creating a secure environment for all citizens." Product liability and federally insured mortgages had already appeared in the second period (and national defense and local poor laws were there from the beginning), but this third period saw a dramatic expansion of federal disaster relief, health, safety, environmental protection, federal insurance and financial guarantees, expanded welfare provisions, and new liability laws. (Moss 1999)

But, as Moss notes, "federal disaster policy represents an intricate patchwork of disparate programs and commitments." (321) This was to be expected since, in contrast to European nations, we had a weak central government (three branches of government checking each other; substantial power to the individual states) that was adverse to both planning and a big role for the federal government. Unplanned, the additions were opportunistic and irreversible. Not only was the obvious constituency of public facilities such as schools included, but disaster spending soon extended to private businesses, given their influence, and even individuals in politically sensitive areas as well. For example, subsidized home loans for disaster victims rose from only eleven in 1953 to 195,762 by 1972, and could be very much larger today. In 1955 the federal government covered only 6.2 percent of total damages from Hurricane Diane; by 1972 it covered 48.3 percent of damages from Tropical Storm Agnes and has continued to cover roughly half of uninsured losses from major disasters since then. (327)

Moss attributes this to rising expectations of the citizens; they began to take federal disaster relief for granted. But this may just as well have been the consequences, rather than the cause, of increased funding. The cause may have been the power of business

lobbies in Congress. Business and local governments saw that there was money to made from disasters, and lobbied, in the patchwork of disparate programs and special interests, to get taxpayers in all of society to cover the losses of the few. The losses of one area are "cross-subsidized," or "socialized," and paid for by everyone. (Wamsley and Schroeder 1996) This is only a sound policy if responsibilities come with the subsidization: insurance that one must take out and pay for, and strict standards that reduce the damage to buildings and limits their location in risky areas. That did not happen. Instead, disaster payments followed a political path.

Roughly fifty disasters a year have been declared since the 1990s. Thomas Garrett and Russell Sobel (2002) conducted a study of disaster declarations and the amount of funds allocated by the Federal Emergency Management Agency (established in 1979). They found that the president's decision to declare a disaster and the amount of funds allocated for the disasters were determined by the political interests of the president and of the twenty or so congressional committees that had oversight over FEMA. The presidents favored the states that were politically important for their election and re-election plans. States having congressional representation on FEMA oversight committees received a larger share of disaster expenditures, and in election years the association increased. Garrett and Sobel's evidence was so strong that they could conclude: "Our models predict that nearly half of all disaster relief is motivated politically rather than by need." (Garrett and Sobel 2002) Organizations provide bundles of resources to be used, and thus are tools. The outrage over politically motivated relief following Katrina was justified by its extent, but it was not new.

Comparatively little of the federal disaster funds from the 1960s to the present went to reducing basic vulnerabilities; instead, more and higher levees and better warning systems were the result. (Better escape routes for evacuations were also funded, and since this is a temporary reduction of the size of the human target, it addresses a basic vulnerability issue to some degree.) Most businesses and most homeowners did not take out federal insurance, even though it was heavily subsidized through taxes on the rest of the

nation. Farmers without insurance, for example, qualified for aid almost as easily as those that bought insurance. A farmer could recoup as much as 90 percent of the value of his crops if the area was declared a disaster area. (Moss 1999, 331) The federal government, from time to time, attempted to put limits on the amount of relief, for example, denying soil-repair grants to anyone involved in three or more disasters over the previous twenty-five years, but Congress blocked the requirement. The failure of the government to credibly force business and citizens to take out insurance and change risky practices generated a "moral hazard" problem: "Citizens who feel confident that the government will pay are more likely to forgo insurance coverage and to live in more hazard-prone areas," writes Moss. (335)

Thus, not until the 1950s was the federal government much involved in large natural disasters. The outrage over its failure in the case of Katrina is quite new; there was no substantial federal response to the decimation of Galveston, Texas, in 1900. That event took eight thousand lives, and there was no expectation of federal relief, let alone federal responsibility for failing to protect the citizens. We have come a long way, but there are inevitably many costs. On the positive side, federal involvement has meant raising construction standards and some regulation of land use in risky areas. But this "preemptive" role is quite secondary to the expenditure of vast sums of money on construction (dams, levees, channels) and on reconstruction, insurance payments, and direct aid. Where there is money, there will be political influence—favored states are more likely to be declared eligible for aid—and the opportunity for corruption; and since organizations are tools, they may be used for ideological ends, too (like rounding up dissidents, as we will see). Federal involvement is essential in a society that continues to generates new opportunities for large disasters as it concentrates its population in risky areas, concentrates hazardous materials, and concentrates private power in our critical infrastructure. Unfortunately, the kind of federal involvement we have had, as we will show in the case of FEMA, has done very little to reduce the concentrations that so endanger us.

## BIRTH OF FEMA

It was in the 1960s and 1970s, with increased prosperity and a rising standard of living, that the impact of natural disasters became more apparent and burdensome. The weather had not changed, but the number of people and structures in the way of its most violent expression had. It was apparent that the federal government needed to play a stronger and more coordinating role. The Department of Housing and Urban Development (HUD) had an agency for disaster assistance, but most of the effort was scattered among more than one hundred federal agencies. The cluster of major hurricanes in 1962, 1965, 1969, and 1972 were not that unusual, but they were now more damaging. Nor were the two big earthquakes in this period—the 1964 Alaskan earthquake and the San Fernando earthquake in Southern California in 1971. They all brought about increased legislation: the 1968 National Flood Insurance Act and the 1974 Disaster Relief Act enlarged the federal role.

But it was the National Governors Association in the late 1970s that prompted a major reorganization. We will meet several professional associations such as this in this book. They often (not always) take a more national and liberal view than most of their members. The professional staff search out issues of concern to their members and move them toward positions that will benefit the community of members, rather than individual members, even the most powerful. One of the major goals of the Governors Association was to decrease the number of federal agencies with which state and local governments were forced to work, in spite of some very cozy relations favored states had with the federal government. Coincidentally, though it may not have been that significant, President Jimmy Carter was nearing the end of a large-scale federal reorganization initiative. Partly in response to the association's devastating report on the fragmentation and politicization of the federal response, an executive order from President Carter in 1979 merged many of the separate disaster-related responsibilities into the Federal Emergency Management Agency. It was heavy lifting. FEMA absorbed the Federal Insurance Administration, the National Fire

Prevention and Control Administration, the National Weather Service Community Preparedness Program, the Federal Preparedness Agency of the General Services Administration, and the Federal Disaster Assistance Administration activities from HUD.

## THE UMBRELLA AGENCY

But there was too much political opposition to creating a consolidated and integrated organization with statutory powers over the merged units, as the Governors' Conference study envisioned. Congressional committees were not about to give up their power over the agencies that would be merged. Committees are where the power of Congress lies, and a committee with few agencies to oversee has less to do and thus less power. So, FEMA became an "umbrella agency," hovering over dozens of programs that retained their identity and retained control of much of their budget. (The Department of Homeland Security, which incorporated FEMA in 2002 is much the same, adding another layer of committee control over the FEMA activities, as discussed below.) This had consequences. One was that the planned integration of programs did not occur. In theory, when a disaster struck, a "virtual" organization would be formed out of the disparate parts, but the integration was less than desired. It would have taken more commitment from the White House than President Carter provided to bring the warring agencies into line.

In addition to the problem of getting autonomous agencies and their congressional committees to cooperate—that is, give up autonomy, which is what cooperation means—there was another serious problem: the communication demands of disaster response. One of the rationales for a new agency was that it could install advanced data and communication systems for handling disasters. Even if the separate agencies desired to cooperate, their communication systems could not, just as the police and fire department's separate systems could not communicate during the 9/11 tragedy. But the integrated communication effort never got off the ground,

and the reason was quite unexpected, and it was the second fateful consequence of the umbrella agency format.

In a move that quickly diverted the agency from natural disaster relief, the notion of disasters was broadened to deal with "all hazards," most notably including deliberate attacks from foreign powers, namely, a nuclear strike by the USSR or its satellites. Moreover, it would deal with the problem of foreign agents fomenting dissent using domestic radicals, and with public protests against, say, a U.S. invasion of war-torn Central America countries. Civil defense responsibilities were transferred to the new agency from the Defense Department's Defense Civil Preparedness Agency. We will return to this move later, since its significance was considerable. But first, a quick look at the prosaic difficulties of delivering decent service when so many political interests must be served.

## THE POLITICAL DUMPING GROUND

With the devastations of California's Loma Prieta earthquake in 1989 and Hurricane Andrew on the East Coast in 1992, major national attention turned toward FEMA. It had not been faring well. During the Reagan administration, which started just after the agency was formed by the Carter administration, it was damaged by scandal, organizational turmoil, and political conflict. The agency was seen as a "dumping ground" for the president's political cronies (Waugh 2000, 29), just as the Department of Homeland Security was to be seen in 2003. This is one way organizations can be of use.

There were structural problems as well. Conceiving of itself not as the nation's 911 responder but as the responder of last resort, after local and state agencies were overwhelmed and called it in, FEMA waited for North Carolina officials to ask for help during Hurricane Hugo, in 1989. But the emergency communication system in the state had broken down—the state officials in charge could not effectively communicate with local officials and could not get damage assessments to document their request for federal

aid from FEMA. FEMA officials waited for the governor's formal request, and the assistance was several days late. Rather than being reactive, FEMA should have been proactive, and have prompted state officials to ask for power generators and other equipment needed, and staged supplies close to the area. (Waugh 2000, 29) Three years later, with George H. W. Bush as president, it was still slow in responding to Hurricane Andrew, and while it did what it was supposed to do (waiting for requests rather than being proactive), the preparations were no match for the category four hurricane (category five is the worst). (The storm was initially predicted to hit Miami, which would have meant an incredible disaster; as it was, it was the nation's worst to that date.) The account of political scientists Gary Wamsley and Aaron Schroeder is worth quoting at length. They wrote it before Katrina, but it appears that nothing has changed since 1992.

> Things rapidly came unraveled. Neither the county nor the state emergency management systems were prepared for the destruction of a Class IV hurricane. Emergency management personnel, police and fire departments, power companies, and others who normally would have been the mainstays of disaster response were victims of Andrew themselves. No one was able to mount an effective assessment of the damage or of medical and life support needs. Officials in the state EOC at Tallahassee kept pleading with local officials to tell them what they needed, and frustrated and equally frantic local officials kept saying they did not know what they needed— "Send Everything!" To which agonized state officials could only reply, "We can't send everything."
>
> Finally on the third day after the hurricane (and three days after a visit by President Bush and the Director of FEMA), the chaos, frustration, and lack of large-scale help from anywhere prompted Kate Hale, the Dade County Director of Emergency Preparedness, to hold a press conference. In exasperation, she uttered a politically explosive sound bite: "Where the hell is the cavalry on this one? We need food. We need

water. We need people. For God's sake, where are they?"
(Wamsley and Schroeder 1996, quoting *Newsweek*, September 7, 1992, p. 23)

The response to this nationally televised cry of anguish was that
FEMA, which had been responding "with what it considered better than average speed," was bypassed by the White House and the
next day 7,000 federal troops arrived in Florida, with more on the
weekend, until almost 20,000 military personnel from all branches
were there, including nineteen generals. In addition, the Secretary
of Transportation was dispatched to take charge, and FEMA was
pushed aside. The authors indicate that extensive interviews with
disaster personnel suggested that the troops, in part, made things
worse—after all 20,000 had to be transported, housed, and fed,
causing congestion. It was the first major participation of federal
troops in a hurricane disaster, and was controversial.

In an extensive research effort, Wamsley and Schroeder concluded that FEMA was cobbled together in 1978 by a rapidly declining presidential reorganization project under President Carter.
It had never fully overcome the results of its many birth defects,
they argued. It has been an agency torn by turf fights along program lines, overburdened with political appointees, stuck with a
"Mr. Bumble" image, labeled the "federal turkey farm" for the
quality of its appointees, tagged with a reputation for petty sleaze
and tacky scandals, seen as trying to play a major role in national
security without the necessary skill or clout, and perceived of by
other agencies as claiming more power to coordinate the rest of the
government than it had muscle or capability to do. (Wamsley and
Schroeder 1996)

The details are convincing, illustrating the difficulties of overcoming our vulnerabilities through normal governmental processes, given the strength of congressional committees and the interests they represent, and the distracted attention of the White House
toward natural disasters that are relatively rare and, when they
occur, can be blamed on God.

Although FEMA was (and still is) relatively small (about 2,200

employees in 1996, when Wamsley and Schroeder were writing), it was forced to take more than thirty political appointees, typically with no disaster-related experience and thus unqualified for the posts, a number exceeded only by the highly politicized Office of Management and Budget. Nine of the these thirty are presidential appointments, requiring log-rolling with the Senate, which must confirm them; this involves six different Senate committees, each with its own program interests. It is small wonder, the researchers commented, that it has been accused of being a dumping ground for political appointees. (Wamsley and Schroeder 1996, 240) Disasters create employment for political supporters. About twenty committees in the House and Senate had legislative jurisdiction over the programs and appropriations of FEMA, and more than *two-thirds* of House and Senate committees could be involved with its activities. No single committee had comprehensive oversight over FEMA's activities, just a myriad of committees and subcommittees with diverse and conflicting interests.

In a 1992 report on internal controls, FEMA formally recognized the problem of fragmented jurisdiction when the director commented:

> FEMA's programs are authorized and directed by a myriad of enabling legislation, appropriations acts, executive orders, and National Security Directives. In addition, congressional oversight and jurisdiction involves some 16 congressional committees and 23 subcommittees. As a result, FEMA's mission is continually altered and shaped in a piecemeal fashion by diverse events, the influence of various constituencies, and differing congressional interests. For FEMA's management, appropriate integration of these various authorities into a cohesive mission is difficult at best, especially given the fragmentation and dynamics of legislative policy. (Wamsley and Schroeder 1996, 241)

Stopping there, and looking at Katrina thirteen years later, we are tempted to say either that nothing has changed or that FEMA's failure in Katrina was no worse. After all, a major point of this book

is that we cannot expect much from our organizations. That is true, but there are circumstances where organizations exceed their expected, normal failure and others where they perform better than we might expect. Much of this variation depends on how they are used. As shown below, FEMA has been at both extremes. FEMA performed poorly in its first year, partly because it was new and partly because President Carter's authority eroded severely with the extreme inflation and the Iran hostage situation that he faced. Its performance worsened under Presidents Reagan and George H. W. Bush as it changed its focus and structure. Then, under President Clinton, it became a model agency and began to consider vulnerability reduction. Under the second President Bush, it has reverted to the low level of operation it had under President Reagan. It is an important story of how federal agencies can be used.

### HIGHJACKING FEMA

The organizational problems of FEMA were only a part of the problem. While Congress and interest groups found ample uses for the agency, its political opportunities, and its money, ideological interests quickly found it to be rich ground for cultivation. It all started innocently enough. When the agency was formed, a civil defense effort was part of its mandate. The charter called for planning and training activities concerning "natural disasters, nuclear war, the possibility of enemy attack on U.S. territory, and incidents involving domestic civil unrest." (Churchill and Wall 2002) FEMA's first director, John Macy, emphasized the similarities between natural hazards preparedness and preparedness for a nuclear attack (now labeled the "all hazards" approach). During its brief history in the Carter administration, however, the civil defense program was clearly subordinate to the emergency relief program; it was just one other possible disaster it would deal with. The Cuban Missile Crisis was long past, and the world recognized our huge superiority in nuclear weapons. But the Carter presidency ended in 1980 with the election of Ronald Reagan, and a year later natural

disasters took a subordinate role to civil defense, even though natural disasters kept coming and there was no appreciable civil defense threat.

In his first year as president, Reagan extensively reorganized the federal government. Then he turned to FEMA in 1981, and appointed Louis Giuffrida, as its new head. Giuffrida had been a lieutenant colonel in the U.S. Army's military police, and after retiring he joined the California National Guard as a general (and insisted on being called that). When he was governor of California, Reagan had chosen him to head up a civilian defense program. One of General Giuffrida's significant initiatives in California was to establish a training camp for antiterrorists and to plan for the detention of large numbers of agitators and aliens. He brought this program into FEMA when he was appointed head of the agency, giving the civil defense program a new status. The agency set up the Civil Security Division and a training center for handling agitators and terrorists for more than one thousand civilian police. It also began gathering files on U.S. activists. The apparent intention was to round these up and place them in internment camps in the event of civil unrest. One national training exercise envisioned incarcerating 100,000 "national security threats." (Churchill and Wall 2002; Reynolds 1990; Ward et al. 2000)

A top secret national security directive (NSDD 26) that Reagan issued in 1982 effectively linked FEMA with the military and the National Security Council (NSC). In FEMA, a small division, the National Preparedness Directorate (NPD), was charged with developing a classified computer and telecommunications network to ensure the continuity of the government in the event of a nuclear attack. The network was developed by the National Security Council and subsumed within the broader national defense information network of the Department of Defense (DOD). Though it was originated by FEMA, and drew on more and more of FEMA's budget, the agency's disaster relief personnel could not have access to it. It was "top secret"; only the DOD and the NSC could access it. Moreover, Congress could not examine the activities or budget of the civil defense part of FEMA. (Ward et al. 2000)

This proved to be a fortunate rabbit hole for our nuclear war defense system to hide in. Since the late 1960s, the public's support for large civil defense programs had waned. The viability and effectiveness of the programs were questioned, and "Civil Defense, as a publicly funded program and profession, faced a bleak, and possibly short, future." (Ward et al. 2000, 1022). Critics charged that the response to natural disasters took a back seat to bomb shelters in homes and mass evacuation plans, always a part of FEMA, though a small part in the past. (We now recognize these plans as ludicrous: see Lee Clarke's [1999] riotous account of government plans in this period to evacuate the East Coast in the event of an attack and Ted Gup's [1994] hilarious account of the foulups during drills for evacuating top federal officials to the Greenbrier, West Virginia, and Camp David redoubts.) Now nuclear defense would be secretly revitalized in FEMA, and FEMA's National Preparedness Directorate would expand substantially.

The result was that the original rationale for establishing FEMA —bringing together the many agencies involved in disasters in a communication network that would allow coordination—was defeated. The communication and data-gathering resources of NPD were state-of-the-art, but they could not be used by the rest of FEMA. In fact, the rest of FEMA could not even get assistance in setting up their own communications system from NPD and its DOD and NSC allies. They had to pursue their efforts on their own, and that meant many incompatible (and technically low-level) systems. "This lack of coordinated IT [information technology] development led to a situation where, by 1992, there were over 100 different IT systems operating within FEMA's other divisions, the majority of which were incompatible in terms of network interfacing." (Ward et al. 2000, 1022) Normally, we would severely criticize FEMA for this incompatibility; knowing that the employees in the field had no choice but to craft their own systems puts it in a different light.

FEMA's NPD was also responsible for building three expensive, state-of-the-art relocation centers (should Washington DC be hit by nuclear weapons), a fleet of four Boeing 747s, and five ground

mobile support and communication systems, with all these assets in constant communication. None could communicate with the rest of FEMA. "The result was that FEMA developed one of the most advanced network systems for disaster response in the world, yet none it was available for use in dealing with civilian natural disasters or emergency management." (1023) (See also Churchill and Wall 2002; Reynolds 1990)

The NPD went further in stymieing the rest of the agency, and it was deliberate, not inadvertent. A General Accounting Office (GAO) report in the second year of the first Reagan administration, 1981, had faulted FEMA's lack of IT resources and recommended computer models and networks for damage assessment, and the NPD division in FEMA responded that this was impossible because each disaster was different! Unbeknownst to the GAO and to Congress, it hired a company to develop just such a disaster assessment program. But when it was completed, it was shelved. "It was felt, by the officials [in NPD] that release of the software and recommendations would undercut the Continuity of Government program" that NPD was developing. This caused FEMA's inspector general to question the priorities of NPD, which the agency ignored. (Ward et al. 2000, 1024) FEMA remained, throughout the 1980s, with a paper-based system, relying on the public telephone network, failings which showed up in 1989 with Hurricane Hugo, prompting U.S. Senator Ernest "Fritz" Hollings, from the devastated state of Florida, to declare that FEMA was "the sorriest bunch of bureaucratic jackasses I've ever known." The next year, when disasters hit his state, Representative Norman Y. Mineta of California declared that FEMA "could screw up a two car parade." (1024)

A report in mid-1992 from a congressional committee criticized FEMA's lack of technological development and lack of support for emergency management, and its civil defense priority over natural disaster response, and labeled the agency a dumping ground for incompetent political appointees. (1025) Two months later, on August 24, Hurricane Andrew hit. With no support from NPD, FEMA had to purchase three hundred personal computers and commer-

cial software locally, train its personnel in using computers (which had been in widespread use in business and government since the 1980s). But the system it set up was so primitive that operators had to take messages in writing and then key them into the computers later, leading to errors, lost reports, and delays in providing assistance. At the end of 1992, "it seemed likely that Congress would either abolish FEMA, or reassign the majority of its programs to other departments of the federal government." (1025) This might have suited the cold warriors handsomely, since their directorate in FEMA could simply go intact to the National Security Council or DOD.

Others involved in the FEMA civil security operation included the controversial General Richard Secord, who was in charge of illegal arms sales to Iran to get the hostages freed, and the even more controversial Lieutenant Colonel Oliver North. North was assigned to FEMA by former National Security Advisor Robert McFarlane, who engineered the "guns for hostages" program wherein the U.S. diplomatic hostages, held by Iran after the Ayatollah Khomeini took over, were freed in exchange for weapons to be used by the contras in Central America. North served in FEMA from 1982 to 1984.

Eventually, the Justice Department concluded that FEMA's activities were openly unconstitutional, and the FBI charged that the head of FEMA, General Giuffrida, had engaged in "de facto nepotism," since he placed only his cronies in key positions in the agency, had improper relations with contractors, and that he had misspent taxpayer monies. This included spending $170,000 to furnish an opulent bachelor apartment in Maryland. (Churchill and Wall 2002; Reynolds 1990) General Giuffrida and his top aides were forced to resign in 1985 after investigations by a House committee, the Department of Justice, and a grand jury. (Giuffrida was apparently not disgraced because of this—he received the Distinguished Alumni Award from the University of Connecticut two years later.) But his departure did not deflect the new thrust of the organization. Organizations are useful. FEMA was the convenient site for the Far Right to mount attacks on the Left. (Or, if that is extreme, to place the threat of nuclear or terrorist attack higher

than the threat of natural disasters.) However, these organizational uses can run afoul of the interests of other organizations, such as the FBI and the Justice Department. With their own security programs being jeopardized by the general, they made an issue of nepotism, graft, and embezzlement, something that would normally not exercise them unduly.


## FEMA'S RESCUE

This diversion of an agency's resources in pursuit of an ideological position about the "evil empire," as Reagan labeled the USSR, illustrates the vulnerability of our disaster response system. It also illustrates the dangers of an easy blaming of the agency on the grounds of incompetence and being a dumping ground. Both charges are true, of course, but the explanation has deeper roots than administrative ineptness and corruption. All but one of the five divisions (directorates) of the agency were denied the resources to undertake its primary task of coordination of agencies involved; the assets were present in the civil defense division, but it was not interested in natural or industrial disasters. The issue is primarily political, not organizational. Congress had its teeth in FEMA for political reasons, but the biggest jaw was the White House.

Wallace Stickney, who headed FEMA during the George H. W. Bush administration, tried to free it up from dominance by the civil defense wing, but the wing was too strong (and possibly the president was too unconcerned). Stickney was blocked by the intelligence community, according to Gup's account, which includes interviews with Stickney. (Gup 1994)

By the time President William Clinton succeeded President George H. W. Bush in 1992, the rationale for a heavy emphasis on the threat of a Soviet attack had long disappeared. In the late 1980s, the Soviet Union had collapsed and its military might was found to have been grossly exaggerated by the CIA. A significant Clinton initiative was the "reinventing government" program, headed up by Vice President Gore, and a keystone of that program

was the "information superhighway," encompassing information technology and networking among organizations. FEMA was an obvious candidate for reform. The GAO issued a report critical of the agency, and Congress asked for a study by the National Academy of Public Administration, which focused on the problem of political appointments. (Waugh 2000) In 1992 Clinton appointed James Lee Witt to reform FEMA, and just as important as this appointment, he made it a cabinet-level agency, reporting directly to the president. Just as Giuffrida had been a buddy of Ronald Reagan who shared his ideology when Reagan was governor of California, so was Witt an Arkansan friend of Bill Clinton when Clinton was governor. The difference was that Witt was a fourteen-year veteran of state emergency management—that is, experienced in FEMA's original mandate. Furthermore, the threat of the evil empire had succumbed to the antiseptic of daylight, and the civil defense program that had hijacked FEMA was even less urgent.

Initially, Witt did not appear to be change-oriented. He opposed a reorganization bill based on recommendations of the National Academy of Public Administration. The bill died in committee. But by November of 1993 he proceeded to reorganize the agency along the lines of the defeated bill and most significantly tackled the information technology problem. The agency got good marks over the next run of disasters, including preparations for a hurricane that did not come ashore, severe Mississippi Valley floods, a California earthquake, floods in Texas and Georgia, and the Oklahoma City bombing.

Writing in 1995, Wamsley and Schroeder were reflective; could the difference be a good executive despite an unwieldy and ill-trained staff? They mused: "Perhaps it should be sobering to think about how much difference a good political executive with roughly 3,000 senior political executives of little skill and experience and an average tenure of two years can make?" (Wamsley and Schroeder 1996, 243) Witt's exceptional skills included a sense not just of what an emergency agency should be about but of how to handle the political interests of Congress. There also were no disasters of

the magnitude of Andrew or Katrina. But I am inclined also to suggest a more critical factor: the White House allowed the agency to have a coherent and legitimate focus and gave its head the authority and autonomy to pursue it. It backed up Witt in his tussles with Congress, for example. The White House can use its agencies for inappropriate or appropriate purposes, or can ignore them, in which case they will either go their own way (the "powerful bureaucracy" example, sometimes so powerful as to countermand White House orders) or be captured by the powerful interests in Congress (the "captive agency" example, such as the Nuclear Regulatory Commission, as we shall see in chapter 5).

William Waugh, in his primer on emergency management (2000), shares the enthusiasm of administrative theorist Charles Wise (Wise and Nader 2002), and Wamsley and Schroeder, for Witt's new leadership. FEMA was one of the biggest success stories of the Clinton administration, Waugh writes in his 2000 volume. Helped by a declining emphasis on civil defense as a result of the end of cold war, Witt emphasized mitigating the effects of natural disasters and more collaboration with state and local government and the private sector, according to Waugh (32) and others. We will mention some of these changes later, but the most important one, a brilliant and detailed account that we will draw upon by Robert Ward, Gary Wamsley, Aaron Schroeder, and David Robins, concerns information technology, or accessing the super secret information highway in FEMA, the National Security Agency, and DOD. (Ward et al. 2000) (For a fuller account of the episode that follows, see Schroeder, Wamsley and Ward [2001].)

### TAKING IT BACK

Witt negotiated with the National Security Council and the Department of Defense and got portions of the information technology assets declassified so that FEMA could use them. These assets included the emergency mobile fleet, computer modeling of damage projections, and satellite surveillance feeds for area damage

assessment, none of which FEMA had been able to use. The agreement also allowed FEMA to link up to the DOD "backbone," the fast and secure major servers that carried its traffic. This allowed computers in the field to have satellite uplinks and downlinks for on-site assessments and real-time audio, video, and data feeds from ground locations. More advanced but classified systems remained under DOD control, and a series of agreements were established for use of these in emergencies, with the Army serving as the agency arranging support. All this helped enormously, but it was still cumbersome to get authorization, and the DOD backbone was unavailable to nonsecure systems at the state and local levels of government. FEMA needed its own network. With the National Preparedness Directorate being dismantled or at least defanged by Witt, its assets were placed in a new directorate, Information Technology Services (ITS) and charged with providing agency-wide services for routine operations and emergencies.

Even before the reorganization, while NPD ruled information technology, parts of FEMA were struggling to develop better tools. A Miami-based geographical information system (GIS) software company contacted FEMA shortly after Hurricane Andrew, and that software eventually provided the agency with a portable, ready-to-go database of information on populated areas vulnerable to natural disasters. Ground structures could be linked to individual addresses and homeowner information, allowing a direct match between structural damage assessments and homeowner assistance applications. Special vans could drive through an area, feeding visual information on damaged buildings into a database through satellite links, with two-way wireless modems sending voice, data, and graphics to field offices. Even before a disaster, since the system could be linked to the hurricane warning system, preliminary assessments of damage that might be sustained could be obtained and the necessary resources mobilized and prepositioned.

A test of a prototype before Hurricane Emily hit the Outer Banks of North Carolina in 1993 estimated the destruction of 674 homes, and the actual damage resulted in 683 claims. Furthermore, FEMA led the way in standardizing the GIS software for all

federal offices that might use it. By 1995 the system was operational and extended to disasters such as floods, earthquakes, fires, and less frequent disasters. All that was lacking was a disaster-response telecommunications network outside the DOD network, which still required cumbersome access. The agency modified a system used by the U.S. Army to provide the nationwide capability that it needed. "The final system provided a 'real-time' environment with direct feed from the disaster area, and all levels of response sharing the same level of information for coordinating the response effort." (Ward et al. 2000, 1028) It was, in many respects, the successful application of what was coming to be called "network-centric warfare," but without the danger of highly centralized micromanagement that the military culture encourages. (Perrow 2004)

Ward and others wrote (perhaps unreasonably) glowingly about the success of this reorganization. Since major disasters involve virtually all of government, the FEMA IT system is linked to virtually all of government. "The system is capable of not only providing a real-time assessment of the disaster impact and magnitude, but also the location of food, supplies, and shelter necessary to sustain human life in the impact zone. Supporting the network are a series of databases and programs dealing with human services, infrastructure support, mitigation, and coordination of emergency disaster response, all available to any level necessary to bring the affected area back into social and physical equilibrium." (1029) But it is unclear why the network failed to work in the Katrina hurricane.

Robert Ward, of Louisiana State University, says in a personal communication that despite the advances of the mid-1990s, the system was not completely integrated; many systems turned out not to be compatible with one another when actually put into action. More important, the system was based on the magnitude of disasters we had experienced. It worked quite well in the four hurricanes that hit Florida (itself unprecedented) in 2004. But it had never been scaled to deal with a response of the size needed in Katrina, and thus programming failures were compounded by a lack

of processing capacity. In addition, the local telecommunications structure in the Katrina area was devastated and the volume of communications overwhelmed the system. (However, one would expect this to have been anticipated.) Backup generators ran out of fuel. The Mobile Emergency Response Systems (MERS) needed to provide the emergency link, and the portable field transmissions units needed to communicate with MERS, were initially pre-positioned in the wrong places, and so on.

However, enough data did get through to fully inform top FEMA, DHS, and White House personnel about the gravity of the situation. As we will see, those levels did not act appropriately on those data. While the system was not designed to handle the widespread devastation of Katrina and then Rita, it appears that the advances made under Witt were slowed when President George W. Bush took over. As we will see in the next chapter, the likes of Ollie North and Richard Secord have reemerged.

## OTHER FEMA INITIATIVES

But before examining what happened to FEMA when it was moved into the new Department of Homeland Security, let us return to some of the other initiatives of the FEMA of the 1990s. They will bring us back to the basic issues of a federal response system: who gains, and who pays.

A major problem has been the credibility of federal threats not to help those who do not take out subsidized insurance or locate and build to federal standards. After the one-two punch of Hurricane Andrew in 1992 and the Northridge earthquake in early 1994, some changes were made. Since private insurers were pulling out of risky hurricane and earthquake areas, the demand for governmental-subsidized insurance mounted. First at the state level, then at the federal level, tougher laws and regulations appeared. At the federal level, incentives for flood-mitigation efforts and tighter enforcement of flood-insurance requirements appeared. The National Flood Insurance Reform Act of 1994 created a community

rating system, giving points for specific mitigation efforts, to determine insurance rates for the community. (Waugh 2000, 173–74) Tighter enforcement meant that no federal disaster relief would be given to those in flood-prone areas that did not have insurance; lending institutions would be penalized for extending loans to borrowers without flood insurance; one could not take out insurance five days before the predicted crest hit, instead a thirty-day limit was required; supplemental disaster appropriations by Congress were made more difficult according to Moss. (1999, 339–41) Moss gives no indication of how successful enforcement of these regulations has been, but he admits, as we would anticipate, that the initiatives "are plagued by serious problems." But it was at least a beginning. These measures not only provide better preparedness, they constitute mechanisms for deconcentrating populations in risky areas, something that we are prone to say cannot be done.

However, state governments want to have "healthy real estate markets," and "there remains a serious question about the credibility of the federal government's threat to withhold assistance from victims lacking federal insurance." (342) A massive insurance organization has been proposed to aid the insurance industry with federal financial backing; unfortunately, if predictably, its rates and underwriting standards would be completely unregulated, opening the door to abuse. (343) But academic insurance experts are optimistic about what "an insurance system with rates based on risk" can do. The University of Pennsylvania insurance expert Howard Kunreuther advocates monetary incentives to reduce risk, fines for noncompliance, tax credits to encourage mitigation programs, enforced building codes and land-use regulation—all wise suggestions and needed. (Kunreuther 1998)

Moss recommends adapting key elements of the French system for handling catastrophic disasters, principally, *open* "cross-subsidization" of insurance through government guarantees. Unfortunately, this would mean that those who live in the less desirable but less risky areas (those far from rivers and ocean views, and far from Los Angeles and San Francisco with their earthquake haz-

ards) would pay, through taxes, the vast majority of costs for the insurance of those in the more desirable but risky areas. The argument is that since this already happens, the point of making it transparent is to make it easier for the federal government to then withhold relief from those who remain uninsured. At the least, those enjoying nice (though risky) areas would be forced to buy insurance, subsidized though it is. This would also reduce the insurance companies' "enormous opportunities for gaming the system and cherry picking risks" that exists now. (Moss 1999, 347) As we will see throughout this volume, the absence of a strong regulatory state greatly increases our vulnerabilities, and does not "free up the market" to find the best solutions.

### CONCLUSION

Relief from major disasters is clearly a federal responsibility, and the main agent from 1980 to the present has been the Federal Emergency Management Agency. But there are interests to be served in responding to large emergencies and in seeking to mitigate their effects through regulation. Starting at the top, the White House has election interests, which distort the response (President Clinton was no exception), and next in line is Congress with its powerful committees overseeing the multifaceted aspect of disasters. Political power makes the largest field unlevel. Then local building and real estate interests resist strong building codes and restrictions on the density of settlements in desirable areas, and local interests (farmers, homeowners, businesses) press for subsidized insurance or even for relief in spite of the failure to take out insurance. Basic vulnerabilities—dense settlements in high-risk areas, concentrations of hazardous substances in vulnerable areas, and so on—are rarely addressed because of political pressures.

With these constraints, it is not surprising that FEMA "could screw up a two car parade," as one congressman had declared. But perhaps the most surprising thing is that with White House backing, an able administrator was able to turn the agency around,

equip it with good technology for the time, and do his job despite the pressures just described. Even after it was moved to the Department of Homeland Security in 2002, when confronted with the unimaginable challenge of four major hurricanes striking Florida within weeks of each other, it seems to have performed well. However, the next chapter shows, its natural disaster response capabilities were losing budget support and experienced personnel, and lacking leadership from either the White House or the Department of Homeland Security which controlled it, it fell apart more completely than even during Hurricane Andrew in 1992.

FEMA is not responsible for addressing our issue of basic vulnerabilities. Unfortunately, no one is. The issue has not emerged, despite all the words spilled about Katrina and other natural disasters. Being basic, these vulnerabilities have a long time horizon and there is no natural constituency with electoral power that could publicize the dangers. But what if FEMA were given a mandate to deal with settlement density, escape routes, building codes, and concentrations of hazardous materials in vulnerable sites? We would need a change in our mindset to make basic vulnerabilities such as the size of cities in risky areas and the amounts of hazardous materials in urban areas as high a priority as rescue and relief. But if a government agency had such a mandate (and had White House support), a government agency might be able to do it. Governments have been successful at comparable missions; think of the Franklin D. Roosevelt administration responding to the Depression, the mobilization of the country in World War II, the Social Security program, or the John F. Kennedy administration's drive to put a man on the moon in ten short years, under budget. Is it unthinkable that one of our political parties would put making America safe by reducing its vulnerabilities as its main goal?

# 4 The Disaster after 9/11

## The Department of Homeland Security and a New FEMA

------------------------------------------

IF FEMA REPRESENTS A promising but flawed attempt to involve the government in natural disasters, the 9/11 crisis presented a challenge to our government that went way beyond those that natural disasters had presented for decades. It also presents a challenge to the thesis of this book.[1] The World Trade towers were large targets that could not be reduced; depopulating New York City is not an option. But terrorists could choose to attack any number of other targets that we are now trying to protect, and these targets could be reduced in size. Regardless of that, this chapter will demonstrate the manifold failures of the agency to protect, mitigate, and even recover from terrorist attacks and other disasters. These failures make the case for vulnerability reduction all the more urgent.

The 9/11 crisis presented a challenge to our government that went way beyond any challenges that natural disasters had presented for decades. The terrorist attack suggested possible future ones that would involve more of society than a natural disaster might, and we had few institutional structures to cope with it. Something far

------------------------------------------

[1] An earlier version of this appeared in Perrow 2006a.

beyond the recombination of disaster agencies that produced FEMA was needed. What we got was a very expensive and slight increase in protection from terrorist attacks, and a marked reduction in protection from industrial and especially natural disasters.

The first thing required was a change in the mental model that our top officials in the White House were using to address terrorist threats to the nation. Years of preoccupation with state-sponsored threats would be hard to set aside, buttressed as they were by many institutional arrangements in the rest of the nation. Second, we needed a new institutional capacity for dealing with the new terrorist threat. The question here was whether this capacity should reside primarily in the executive branch (White House) or in the legislative branch (Congress). Without a convincing change in its threat model, the White House, I will argue, was unable to mobilize support for allowing it to control the effort, and control passed to Congress. But such a huge project affected so many interests that in congressional hands, the project was bound to be unwieldy. Furthermore, the institutional framework chosen for protecting homeland security followed a cultural script that organizational designers most easily revert to, namely, centralized control, even though the problem would be more amenable to the empowerment of diverse, decentralized units with central coordination rather than central control. The many organizational difficulties we will examine that flowed from the central control format, coupled with distracting wars and lack of White House urgency, make the failure of the reorganization inevitable.

Finally, the same format was used to attempt a restructuring of the intelligence agencies and once again, powerful interests, this time in the Pentagon as well as Congress, appear to have thwarted this effort as well. Compared with the restructuring response to the 9/11 tragedy, the initial creation and shaky evolution of FEMA, for all its problems, looks like a piece of cake.

Terrorism, of the Al Qaeda variety, has special interest for those who would protect our nation by reducing its vulnerabilities. Unlike natural and industrial threats, terrorism will *intentionally* seek out and strike at concentrations of populations, concentrations of

hazardous materials, and concentrated nodes in our critical infrastructure. I doubt that we will ever eliminate the threat of terrorism, so reducing those concentrations seems at least as important as catching terrorists (since there are replacement terrorists standing in line) or locking the many doors to our way of life.

## TERRORIST THREATS COMPARED WITH NATURAL AND INDUSTRIAL DISASTERS

Deliberate disasters can be directed toward our key vulnerabilities. In contrast to nature and industrial accidents, terrorists are a *strategic enemy,* looking for our weakest and most consequential targets. There is evidence that Al Qaeda has considered targeting nuclear power plants and chemical plants, national monuments with concentrated populations about (the White House, prominent buildings), key government agencies such as the CIA headquarters, prominent infrastructure institutions such as financial institutions, and the heart of our communications system, the Internet.

The *scope* of a terrorist attack could exceed that of most natural disasters in several respects. It may not be confined to a region or a single floodplain or earthquake zone. An earthquake could destroy the Los Angeles airport, but airports elsewhere are not thereby threatened. A terrorist attack on the airport would prompt us to drastically curtail operations at all large airports. (A colleague, Jake Shapiro, imagines terrorists at three major airports going through the terminal with a small rag soaked in an explosive the detectors are set to detect, such as fertilizer, and brushing up against suitcases, briefcases, and coats.) Except for epidemics, natural disasters are limited to local areas; they are not nationwide, whereas terrorism potentially is. The scope is also greater because of the psychological impact; we fear, but generally understand, the industrial or natural disaster, but the terrorist attack is almost beyond our comprehension, increasing the level of tension.

The terrorist threat is not confined to a particular area of our social life, such as our defense policy, or the agencies that deal with

unemployment, transportation, or energy. These are reasonably circumscribed with limited constituencies, so the Departments of Defense, Labor, Transportation, and Energy are agencies that are manageable. The threat involves many more agencies than a natural disaster might, entailing more jurisdictional disputes in Congress than FEMA ever did.

It is true that natural disasters involve many segments of life: transportation, health, buildings and homes, industry, local government, public order, and so on. But to all these, preventing terrorism adds the detection element—information demands that go far beyond the weather forecast—and the deception element, wherein the threat is disguised. We need much more surveillance with terrorism to detect plots, and because of deception, much more alertness, training, and preparation for a variety of scenarios. As elaborate as FEMA's communication system was, as we saw in chapter 3, its system would be no match for the challenge of terrorism. Even our intelligence agencies, which are much better equipped for communication than our disaster response agencies, do not appear to be able to interpret and share information on terrorists, at least judging from their performance in connection with the 9/11 attack. Finally, the nation can emotionally come to terms with natural disasters as "acts of God," mistaken as that is, and regard industrial disasters as inevitable but infrequent "errors." But the willful, intended destruction of noncombatants may bring about a rage that interferes with rational response.

Truly, the terrorist threat goes beyond all others, natural and industrial, and we cannot expect our organizations or our political culture to be up to it. Since 9/11, we hope our efforts have reduced the danger of a terrorist attack to some extent, even if probably not by much. But the safety of the United States has little to do with locking doors or watching mosques. The two main theories of radical Islamic terrorists are that (1) they want foreign troops out of their lands (Pape 2005), and (2) they seek to topple corrupt, authoritarian Muslim regimes such as that of Saudi Arabia, which the United States continues to support. (Benjamin and Simon 2002) According to these analyses, our foreign policy, by putting us at

risk, has far more to do with our domestic safety than anything the Department of Homeland Security can do.

But even if these theories are correct, we have reason to examine our DHS experience very closely. First, the department is concerned with all hazards, not just terrorism, and the five years since 9/11 have seen deaths and destruction from natural and industrial disasters that exceeds the toll of decades of terrorist activity—by domestic terrorists as well as foreign ones. And while a significant terrorist attack may never again occur, we may be certain that our future holds more floods, hurricanes, earthquakes, forest fires, and industrial disasters. Second, our DHS experience is a cautionary tale that exposes some basic fault lines in our political system that must be taken into account as we continue to shape our security institutions. Third, it illustrates the inevitable imperfections of complex organizations, indicating the limits of what we can accomplish with them. And finally, it once again points toward the advantages of reducing basic vulnerabilities rather than just locking doors and setting alarms.

## THE THREAT MODEL

Why was the administration of George W. Bush so unprepared for a terrorist attack on our homeland? The answer to this question is one part of the explanation for the dismal performance of the Department of Homeland Security. (The other part of the explanation is that we expect too much of our organizations.) It is a truism that the military is always prepared to fight the last war. Given the difficulty of predicting which adversary will strike next and how they will strike, this may be the best the military can do. When we live for decades under the threat of foreign attack, all our major institutions, not just the military and diplomatic corps, become firmly configured to meet this threat. The political area develops scripts and slogans to mobilize defense; a candidate or party cannot be weak on defense. Parts of business and industry thrive on defense expenditures and gain more power than those parts that

are hurt by defense expenditures. Research money goes to the organizations that can build bigger bombs or bigger bombers, not to those who develop better body armor, rifles, battlefield communications. The media builds appropriate images of heroes and villains; the judicial system and social institutions such as education are altered to reflect the worldview. (Early in the cold war, university Russian Studies programs were quickly staffed with faculty without PhDs, contrary to university policy; Black Studies programs had a hard time getting started when they came along because they recruited faculty without PhDs, which was considered contrary to university policy, conveniently now evoked.)

A cold war that ran for some four decades can build up an impressive institutional support system that is confident that the next engagement will be similar to those anticipated for the last four decades. Many ways of life are elaborately, and sometimes comfortably, built around this sensible hypothesis.

But it was beginning to erode in the 1990s when no obvious enemy nation was apparent after the collapse of the Soviet Union. Since institutions, by their nature, change slowly and reluctantly, there remained a physical, organizational, and ideological infrastructure that could be mobilized by any president who argued that we remained threatened by foreign states.

Even events that challenged the lingering worldview need not shake it. The United States was repeatedly attacked by terrorists with no apparent state support in the 1980s and 1990s, sometimes on its own homeland. By the mid-1980s, Al Qaeda, a terrorist organization not supported by a foreign state, was identified as a serious threat to our security at home. The Soviet Union had collapsed in 1989, and there was irrefutable evidence that the United States was the only superpower left. But this prompted only minor adjustment in our defense strategy (more rapid-deployment troops; some reduction in nuclear weapons). But our intelligence agencies and military assets still thought of nations as the enemy, even after the first Twin Towers attack in 1993 was laid at the door of terrorists that were not sent by a foreign state.

Regardless of our superpower status, the mindset and the asso-

ciated institutional direction led the George W. Bush administration to place something as difficult and expensive as an expansion of a missile defense system on a high-priority list when entering office in 2000, while counterterrorism funds appropriated under President Clinton were actually cut. This was not a case of institutional lethargy or drift, or even maintaining a steady state. Compared with the two previous administrations, the new administration vigorously reinstated and refurbished the subsiding cold war ideology. It was a trajectory, built on a mindset that had powerful institutional traditions. These can be mobilized. (For evidence of the cold war ideology, see former CIA bin Laden unit head, Michael Scheuer's *Imperial Hubris*. [Anonymous 2004])

Our vulnerabilities, in the administration's eyes, were not crowded skyscrapers symbolizing our economic power, like the Twin Towers, or the heart of the imperialist beast, like the Pentagon (attacked in 9/11) and the White House (almost attacked). The administration saw our vulnerabilities as the thousands of miles of open skies through which a nuclear missile could fly from a ship at sea that perhaps North Korea could someday launch. This vision is also a way of not seeing.

## FAILED WARNINGS

To understand how the Department of Homeland Security and the reorganization of the intelligence community might turn out to be enormous failures, we have to examine the low priority that the White House put on its "war on terror." Before 9/11 we had been repeatedly attacked abroad (in coordinated attacks on two of our African embassies, our base in Saudi Arabia, and the U.S.S. *Cole*), and at home (in the first attack on the Twin Towers in New York City). All were identified as attacks by nonstate terrorists. But the Bush administration dismissed the warnings about nonstate terrorists by the Clinton administration's transition teams on coming into office and the repeated warnings from CIA head George Tenet and security experts such as Richard Clarke and

Rand Beers. It never even convened its task force on counterterrorism before 9/11, despite repeated warnings that Al Qaeda
could be planning to hijack airplanes and use them as missiles.
(Greenstein 2003)

In contrast, the Clinton administration got word in 1998 of a
planned attack in the United States that entailed hijacking aircraft
and then demanding the release of an imprisoned Muslim cleric. It
responded by placing all major airports on the east coast, especially those serving New York, on alert until the end of January.
Steps included intensified passenger screening and intensified supervision of screeners. The next year, 1999, fearing a "millennium attack" (on the basis of much less evidence of terrorist activity than
the Bush White House had in the months prior to 9/11), President
Clinton shared his presidential daily briefings with up to twenty-
five people. (President Bush limited it to six after 9/11.) Clinton activated resources abroad, foiling some attacks, and activated resources at home. Airlines and airports were kept on alert, the
border guards were alerted, and one terrorist was apprehended
who was linked to Al Qaeda. (Perrow 2005; Staff 2004, 127, 533,
n. 2) At least some dots were connected as a result of attention
from the Clinton White House. Similar actions in September 2001
might have foiled the attacks of 9/11, but the even more urgent
warnings were dismissed by the Bush White House.

Just before 9/11, the FBI's requests to add hundreds more counterintelligence agents were rejected (Hirsh and Isikoff 2002) and a
program designed to provide equipment and training for first responders in the event of a terrorist attack was cut. (Van Natta and
Johnson 2002) The report of the Commission on 9/11 (Staff 2004)
notes that the White House tried to cut funding for counterterrorism grants authorized by Congress, terminated a highly classified
program to monitor Al Qaeda suspects in the United States, and in
general downplayed the threat. In the early days *after* 9/11, the
White House cut another FBI request for counterterrorism funds
by nearly two-thirds, its requests for computer networking and
foreign-language intercepts by half, a cyber-security request by
three-quarters, and a program for increased collaboration entirely.

(Milbank 2006; 2005; for other examples see Perrow 2006a) The major achievements of the administration in the months after the attack were the invasion of Afghanistan to destroy bin Laden's base and training ground and speeded-up preparations for an invasion of Iraq. Both were conventional operations, whereas 9/11 also required unconventional ones. Even the Afghanistan operation barely pursued bin Laden; instead it "outsourced" the job to unreliable Afghani forces. The domestic terrorism program was limited to a small office of fifty or so professionals in the White House to deal with homeland security, set up shortly after 9/11.

## ORGANIZING OUR DEFENSE

### The First Homeland Defense Option: White House Control

Political scientist Charles Wise usefully lays out three options available to the government to cope with the threat to homeland security. The threat was acknowledged years before 9/11, of course, and the three options had been considered in various governmental reviews and task forces. Wise labels the option first undertaken by President Bush in the month after the attack as "executive order coordination." (Wise 2002a)

This option was to coordinate and stimulate homeland defense from the White House by establishing an Office of Homeland Security (OHS), and appointing an assistant to the president to run it (Tom Ridge, the former governor of Pennsylvania, who was widely liked but had no terrorism experience). While this might suggest a minimalist response, as compared with the new Department of Homeland Security that we eventually got, it was in many respects the most promising option, since it emphasized coordination rather than centralization. It has the advantages of rapid response and flexibility and would give the president more direct authority than he would have with a Department of Homeland

Security, which would give much control to Congress. But while it was in place, from October 2001 to July 2002, the OHS had an uncertain status and achieved little. As noted, the White House was cutting terrorist-related funds a month after 9/11.

This option, a White House office, would require exceptional activism from the White House, whereas a department would not. The presidential Assistant for Homeland Security would have to be aggressive, rather than just likable, and have international experience rather than gubernatorial experience. She or he would have to be strongly supported by the president in struggles with the Office of Management and Budget to reallocate funds and in struggles with the all-important intelligence agencies, particularly the CIA and the FBI, who were not sharing information. For example, using the threat of personnel replacements and budget cuts, the assistant would have to demand concrete evidence that the FBI changed it priorities from catching and prosecuting criminals and especially the purveyors of illicit drugs (its top priority), to investigating terrorist activity and sharing information with the CIA.

It is true that there are laws on the books governing cabinet bureaus and congressional actions that prevent the president or the head of the Office of Homeland Security from running roughshod over them, but presidents have been able to do a great deal outside of them. (Sylves and Cumming 2004) The theme of the "imperial presidency" mobilizing political and economic resources is a strong one in our history, even without crises. In a thoughtful personal communication, political scientist Patrick Roberts points out that while a White House czar for domestic policy can work during a time of crisis, coordinating domestic policy from the White House is more difficult than coordinating foreign policy. Domestic agencies are so closely tied to corporate, professional, and electoral interests that a White House adviser rarely has enough control to coordinate their efforts. Foreign policy is somewhat more removed from electoral politics by its nature and because of congressional committee control, so it is more responsive to White House direction. President Franklin Roosevelt was able to control domestic policy during World War II through a White House aide; Presi-

dents Reagan and George H. W. Bush both tried it but had limited success. (Roberts 2003)

Executive order coordination has the virtue of preserving the decentralized structure of the many agencies that would be involved while adding a new measure of direction that would promote coordination. Decentralization does not necessarily mean fragmentation. Fragmentation comes when decentralized units do not receive explicit policy direction and the general oversight that indicates the policy is being carried out, and do not share information. The overwhelming problem with our homeland security and defense was a failure to coordinate—to connect the dots, as we endlessly heard. Coordination can be achieved through centralized control in small and moderately sized and homogeneous agencies, but mammoth projects are almost impossible to coordinate through centralization. Mammoth agencies require a great deal of decentralization because of the diverse tasks and skills involved. Decentralized systems are coordinated not by giving central orders but by signaling the policy direction, making sure that information is shared, and monitoring outcomes rather than behavior.

What, concretely, would this mean in terms of executive action? I am sure that the FBI director, Robert Muller, who was appointed just prior to 9/11 and was by all accounts more motivated and effective than any of his recent predecessors, can testify that changing an organization's culture is a tough job. But that is what administrators are paid for. Unfortunately, there was no great reallocation of personnel from drug interdiction and prosecution to, say, translating intercepts and documents, huge amounts of which remained untranslated more than three years later. As mentioned above, the White House was cutting these funds, so the FBI was getting signals that terrorism was not much of a priority. But suppose it was a high priority, and the receipt of new funds signaled that? The Assistant for Homeland Security could have ensured that the FBI's priorities were adjusted accordingly. The assistant could also have demanded that personnel be rotated between the FBI and the CIA. Such rotation had been ordered during the Clinton administration, but no one followed up so it did not happen—a typ-

ical Clinton administration failure. Following up would be the job of the Office of Homeland Security, and would not have required a platoon of staffers, only the support of the president.

For example, the FBI's top management would have had to demonstrate that they took seriously and investigated the complaints of FBI translator Sibel Edmonds. She wrote superiors that she was ordered to slow up her productivity in order to justify a bigger budget request for the translation department, and that favoritism had led to hiring a translator who had not only failed to translate messages in Arabic languages accurately but was married to a man who was on the terrorist watch list. (They both left the United States immediately after the 9/11 attack.) The FBI management dismissed her charges, and instead they promoted her supervisor and his two superiors and fired Edmonds. (Perrow 2005) Strong signals from the Assistant for Homeland Security would have made it clear that investigating such complaints, and there were many, would be rewarded, instead of promoting the individuals charged.

Or, to take another illustration of the type of effort required, the so-called firewall erected in 1995 that supposedly restricted information sharing between the FBI and the CIA. The report of the 9/11 Commission notes this was the result of the "misinterpretation" of judicial rulings and procedures. The FBI conveniently saw them as prohibiting information sharing, and the Clinton administration did not address the problem, even though they were warned about it in 1999. (Still, in two crises in 1998 and 1999, information was shared in the Clinton administration.) But the Bush administration received strong warnings in 2000 and 2001 that intelligence agencies were not sharing data, and that the intent of the 1995 procedures was ignored routinely. (Perrow 2005; Staff 2004, 79) These warnings came at a time of higher threat levels than the previous administration had experienced, but still had no effect on the Bush administration. (Moreover, the forty or so warnings in 2000 about Al Qaeda in the presidential daily briefings—accorded a much higher threat level than in the 1998 and 1999 crises—appeared to have no effect.) These warnings of misinterpretations could have been addressed by a homeland security assistant who

was fully empowered by the president and his top staff. It was front-page news that the FBI was not changing its procedures. In fact, the FBI's several attempts to prosecute suspected terrorists were thrown out of court because of faulty procedures. Redirecting the FBI was certainly not a "slam dunk," but it was far from impossible. When given tasks that fall within its capabilities and tradition, it could perform very well. Investigating the first World Trade Center bombing in 1993 and prosecuting the case was an example. Investigating an actual crime that supplies physical evidence and prosecuting the case is what the agency is about. It is not very good at dealing with the shadowy threats of terrorists, long-term surveillance, terrorists's national and international movements, and cooperating with other agencies. This clearly could be changed. FEMA has been redirected at least three times by various White Houses (though it is a much smaller and weaker agency than the FBI). (An enlightening analysis of the difficulty of changing the intelligence community is by Amy Zegart, though she downplays political interests and especially presidential intent; see Zegart 2007.)

A former White House staffer who had been transferred to the new department indicated that she had even less power in the DHS than when she was in the White House. "You had a platform at the White House. Whenever you called a meeting at the White House, the other agencies came," she said. "Now we're over at the department and the agencies didn't come; they came up with all sorts of excuses." (Glasser and Grunwald 2005) This neatly captures the difference between executive coordination and the other options.

In his dramatic memoir *Against All Enemies* (2004), Richard Clarke indicates that President Bush's initial plan might have worked. It envisioned Tom Ridge heading up a White House Homeland Security staff of about fifty professionals to lead, coordinate, and conduct oversight on the many federal programs involved with security and disasters. As Assistant to the President for Homeland Security, Ridge expected to have real authority, but soon complained that everything had to be cleared with the White House Chief of Staff. (Presumably that meant that homeland security had to be politically vetted, making it less than the top prior-

ity of the president.) While this was still consistent with presidential control of the program, it revealed that the president did not intend to take the domestic security issue seriously enough to deal directly with Ridge.

This was not a good sign, but the Office of Homeland Security still might have worked had the president given it a high priority. Clarke observes: "I believe that adept White House coordination and leadership could get the many agencies all working on components of a consistent overall program." It would be difficult, of course, but over the years his own organization, the National Security Council, had managed to be quite effective during the Clinton administration. He continues: "the alternative method, rewiring the organizational boxes, would make us less able to deal with domestic security and preparedness for years to come." (Clarke 2004, 250) He appears to have been correct. It had taken years for the mergers that created the Energy Department and the Transportation Department to jell, and both were smaller. Ridge apparently agreed with Clarke that the last thing needed was the reorganization involved in creating a new department.

Executive order coordination was the option that President Bush maintained without much enthusiasm from October 2001 until July 2002. The lack of enthusiasm is important; it opened the door to other interests.

### The Second Option: Power Sharing

In the second option, what Wise calls the "statutory coordinator" option, the Homeland Security agency is established by law, rather than by presidential order. Laws are passed by Congress, so this gives Congress more power than under the first option, and thus reduces that of the president. This was proposed by the Gilmore Commission, which was considering these matters well before 9/11, having been formed in 1999. (Its official government title was the Advisory Panel to Assess Domestic Response Capabilities for Terrorism Involving Weapons of Mass Destruction.) Despite

commissions like this before 9/11, aside from a counterterrorism group in the White House, little was done in the way of government reorganization. The Gilmore Commission had examined all efforts in its second annual report in December 2000 and concluded: "The organization of the Federal government's programs for combating terrorism is fragmented, uncoordinated, and politically unaccountable." The problem, it said, was lack of central authority, a routine and easy finding that commissions are prone to make. "The lack of a national strategy is inextricably linked to the fact that no entity has the authority to direct all of the entities that may be engaged." (Wise 2002a) It called for a national office for combating terrorism that would have legal—that is, statutory—authority and be located in the executive office of the president. This location gives the president considerable authority, but it is shared with Congress, which writes the laws governing it.

The commission found that oversight responsibilities for combating terrorism presently lay in the hands of at least eleven full committees in the Senate, plus numerous subcommittees, and fourteen full committees in the House, with all their subcommittees, eighty-eight in all, each carefully guarding their turf. (Ornstein 2003) It was the same problem that earlier commissions had identified for FEMA. The homeland security effort had to be centralized, they, and almost everyone else, argued. This would mean new laws, and Congress would pass them, giving it influence. But its influence would not be scattered over eighty-eight committees and subcommittees. The key to the Gilmore Commission plan was that it recommended a reorganization of Congress itself to centralize control over homeland security matters by having a powerful oversight committee in each congressional house. These could override the authority of the eighty-eight parochial and fragmented committees. This would mean that the committees overseeing, say, the Coast Guard or border control, could see their authority reduced. This recommendation never succeeded. Individual committee and subcommittee chairs—which is where the power lies—were unlikely to support this reduction in their power.

But we have to be careful here. I will make much of congres-

sional resistance to giving up committee and subcommittee power. But these committees are closest to the public, and are expected to be responsive to their constituencies. They are responsible for the fact that our federal system is very decentralized (or fragmented, depending on whose ox is being gored). The committees overseeing, say, border control and immigration or the Coast Guard, would be able to cry out that the agencies they oversee are underfunded, or the congressional committee could publicize the failures of the FBI or the CIA to coordinate with the agency the committee oversaw (the agency itself could not) or investigate charges of torture as members of a military affairs committee, or charges of waste in Pentagon contracts, and so on. The agency being examined is unlikely to bring such matters to the attention of the committee or the White House, nor, if it believes it is underfunded, does it have a spokesperson. There is a legitimate congressional advocacy and representative function, as well as funding pork-barrel projects. If a supercommittee was established, the charges of failures to coordinate, of torture, or of underfunding might not surface; any actions a regular committee might want to take could easily be overridden by the supercommittee. Congressional committees and subcommittees have reasons to want to keep jurisdiction and to exercise power, and it is hard to distinguish turf from representation of the interests of constituents. One is tempted to say that if the task is extremely critical—and combating terrorism seemed to be such—the risk of centralizing power in the supercommittees needs to be run.

(The committee or supercommittee members might fail in quite other ways. They could easily put narrow local interests over those of the nation or could be so indebted to large campaign donors as to fail to represent all segments of their constituency. But this is quite another matter. These narrow interests would be present in the two large oversight committees the Gilmore Commission recommended, or, if there were no supercommittee, in the eighty-eight committees. They reflect a problem with our form of government that goes beyond that of restructuring agencies. I will discuss this in the final chapter.)

### The Third Option: Keeping Congressional Control

It was the third option that prevailed—congressional control of a new department with Congress keeping its existing budgetary and personnel and mission control of the agencies and programs thrown into the new department. Based on the Hart-Rudman Commission findings of 2001, it preserved much of congressional committee power. (Gary Hart had been a Democratic senator from Colorado and Warren Rudman had been a Republican senator from New Hampshire, so the interests of Congress were likely to be well considered.) A new cabinet-level department would be created. The power of the president would consist in recommending to the Senate a nominee to be the secretary of the department, and exercising the normal amount of control the president has over any cabinet head, such as for agriculture or defense. While considerable, of course, presidential power would be less than in the first or the second option. Congress is the clear winner.

Political scientist Donald F. Kettl put it as follows, referring to Tom Ridge:

> What the members of Congress left unsaid was that if Ridge remained a presidential appointee without congressional confirmation, they would have little control over his operations. If they could pass legislation authorizing the office, setting out its powers, gaining the right to confirm him in office, and controlling the office's budget, they could dramatically shift the balance of power. Many members of Congress saw this as one of the biggest new initiatives in decades, and they wanted to ensure that they could control its direction. Bush turned them down, saying through a spokesman that the president did not need congressional action to do what was required. (Kettl 2004, 44)

Many agencies would be moved from their previous departmental location to the new Department of Homeland Security, but the agencies would still be under the supervision of Congress. Thus, the Secretary of the DHS would not have the power to use presidential authority to establish new units (e.g., a counterterrorism threat center that would not be under the control of Congress) or

to force coordination through threatened budget cuts of, say, the immigration services, that the first option provided. All the congressional committees and subcommittees would still have considerable say over agency budgets and line authority; no supercommittees would override them, and Congress made sure none with any real power would ever exist.

Nevertheless, even this option created a high degree of conflict within Congress as it considered the new department. One senator was quoted as saying, "Hell hath no fury like a committee chairman whose jurisdiction has been taken away," and the reshuffling would alter some jurisdictions. (Wise 2002b) Congress would lose some budgetary authority, and the power to make some line appointments, but not as much as under the other two options. A bill to establish the DHS was introduced by two senators, with Connecticut's Joseph Lieberman taking the lead. (Democrats still had a majority in the Senate but Lieberman's power was limited to heading a committee concerned with governmental organization that was, until he seized the initiative, of minor importance.) The president eventually reluctantly gave in and introduced a similar bill of his own.

If the president opposed the Lieberman bill, why did he give in after a few months and propose his own bill, which was very similar except that it included even more agencies than the Lieberman bill had? Clarke argues that the White House was about to have two disasters: (1) an unmanageable department that both houses of Congress strongly supported and (2) "the major new piece of legislation in response to September 11 would be named after the man whom the majority of voters had wanted to be Vice President just twenty months earlier." It was better to have one of those two outcomes rather than both, so the President sent up a bill that would be called the Homeland Security Act, not the Lieberman Act. (Clarke 2004, 250)

The president's bill was drafted in secrecy by five staffers unfamiliar with many of the programs they were moving, and "some of the decisions were made almost at random," according to a stunning review of the department in December 2005 by *Washington Post* staff writers Susan Glasser and Michael Grunwald. The plan

caught many agencies by surprise when it was revealed to the cabinet, and behind-the-scenes campaigns by agencies began. Powerful agencies such as the Justice Department and the intelligence community (FBI, CIA, NSA, etc.) refused to allow offices or staff to be moved out of their agencies. Health and Human Services did not want to lose the national drug stockpile. Energy blocked the transfer of the its Livermore lab (atomic weapons), and it made no sense to transfer the lab's radiological detection teams when it was pointed out that the teams consisted of employees with regular jobs who mobilized only during emergencies. A key facility in a large agency is a policy office, but the Budget Bureau vetoed putting one in the DHS on the grounds that we should not foster big government. What policy there was came from the White House's Homeland Security Council, which micromanaged the details of the DHS, asking for regular updates about uniforms for border guards, the curriculum for teaching border inspections, the selection of a single firearm for DHS training academies, and the like. The powerful Bush adviser Karl Rove vetoed a plan by Ridge to secure large chemical plants because it would give new regulatory power to the Environmental Protection Agency and was opposed by the chemical industry. "We have a similar set of concerns," Rove wrote to the president of BP Amoco Chemical Co. Everyone, it seems, foresaw a disaster; one aide to a cabinet member said, "we all expected an ineffectual behemoth, and that's what we got." (Glasser and Grunwald 2005)

But what they got included extensive congressional oversight, written into the new law, and this added to the ineffectualness of the behemoth. I have characterized this option as "keeping congressional control, " but it is more complicated than that. As a colleague at the Stanford Law School, Laura Donohue, points out, in many ways Congress did not achieve new control, but only maintained control over already existing agency functions and does not control many new initiatives. The executive office has substantial resources. The consequential actions related to homeland security were undertaken by executive agencies, rather than ones controlled by Congress. The Department of Justice expanded the Patriot Act without effective oversight by Congress; the intelligence

agencies were unchanged; and the executive branch expanded the national security letters, Foreign Intelligence Surveillance Act warrants, and various collection powers, all of which have been used extensively. Congress dismantled the controversial Total Information Awareness program—intended to turn truck drivers, postal workers, TV cable installers, and others into counterintelligence agents—but the White House continued the course with its Highway Watch, Marine Watch, Neighborhood Watch, and other programs. Though defunded by Congress, the Defense Department's Defense Advanced Research Projects program with its eighteen data-mining operations were transferred to the National Security Agency, CIA, and FBI. (Donohue 2006; Harris 2006) The NSA's data-mining operation on millions of U.S. citizens with the help of all but one of the major telephone companies was a major political issue in 2006.

Thus the question of who won and who lost—the executive branch or Congress—is not as clear-cut as one might think. But Congress, as we shall see, kept substantial access to the barrels of pork that homeland security disgorged.

## EXECUTING OUR DEFENSE

### A Rough Start and Poor Reviews

Given the ignominious birth of the department, we should not expect it to achieve much homeland security for at least a few years, and the initial record sadly confirms that expectation. Even if it had strong presidential backing, the difficulty of merging diverse tasks, funding the new responsibilities, coping with congressional interests and with the inevitable uses to which organizations can be put severely limited the effectiveness of the department.

For starters, the launch was rough and premature. President Bush had resisted congressional efforts to establish the department, but once Congress passed the law, he set an unreasonably ambitious four-month deadline to open its doors to twenty-two agencies. It

had a hard time finding any doors to open and initially, befitting the lack of enthusiasm on the part of the White House, was to be placed far out in the Washington suburbs. Ridge, appointed head of the agency, successfully fought that, but the agency still ended up stuck in the basement of a Navy building several miles from federal Washington without room to house personnel that were to be transferred there. It received little help in staffing; the secretary's staff was very sparse; for weeks some offices lacked phones; the budget was so small that finding funds was a constant preoccupation. Touted as receiving a $40 billion allocation, it received far less in new money. One-third went to other agencies such as the Pentagon and most of the other $27 billion was not new money. Five of the twenty-two agencies had a total budget of $19 billion, which they brought with them, and was counted in the $40 billion figure. (See Matthew Brzezinski's scathing and disheartening details on the failings of the department: Brzezinski 2004a; 2004b.) Congress's dozens of committees still have oversight claims on the department, through their old ties to the agencies.

In such a situation, strong leadership on the part of Tom Ridge may not have been enough to rescue the failing launch. Still, according to the research of Susan Glasser and Michael Grunwald (2005) it was not forthcoming. Ridge did not insist on including a policy shop and spent a good deal of time worrying about the image of the department. Organizations such as the DHS should be designed to work with an average leader; we cannot count on getting an extraordinary one. We didn't.

Glasser and Grunwald conclude, in their review:

> To some extent, the department was set up to fail. It was assigned the awesome responsibility of defending the homeland without the investigative, intelligence and military powers of the FBI, CIA and the Pentagon; it was also repeatedly undermined by the White House that initially opposed its creation. But the department has also struggled to execute even seemingly basic tasks, such as prioritizing America's most critical infrastructure. (Glasser and Grunwald 2005)

Congress has watched the agency very carefully. Secretary Ridge said that in the first year, he and his top assistants testified 160 times, about every day and a half, before congressional committees, and counting staff in general there were more than 1,300 briefings on the Hill. It takes from twenty-four to forty-eight hours to prepare for a briefing, he said. And this did not count the "hundreds and hundreds" of Government Accountability Office inquiries. (U.S. Congress 2004) The executive coordination option may have had fewer resources than a department, but they could have been better focused.

The agency has been watched carefully by several public interest groups such as the Council on Foreign Relations, the Heritage Foundation, and the Brookings Institution; by quasi-government groups such as the Rand Corporation and the Center for Strategic and International Studies; and most carefully by the Government Accountability Office (formerly the General Accounting Office), which is asked to do studies for Congress. There are some differences among the reports. For example, the Heritage Foundation emphasizes surveillance of citizens, and recommends that local law enforcement personnel "submit annual assessments of the events, activities, or changes in demographics or patterns of behavior of groups in their jurisdiction," as a Rand Corporation summary of recommendations by commissions and public interest groups notes. (Parachini, Davis, and Liston 2003, 17) (This would reinstall Admiral John Poindexter's Total Information Awareness program in the DHS.) The Brookings Institution, in contrast, has an economic tilt, and is the only one to weigh the budgetary implications of its own recommendations. It is easy to recommend actions as if they were costless.

The previous recommendations of the public interest agencies, like those of the dozens of commissions (Bremer, Hart-Rudman, Gilmore, etc.), suffer from heavy generalizations that evade the tough issues, urge more spending than is feasible, and urge actions without any guidance on setting priorities, as the Rand Corporation summary notes with sadness. (Parachini, Davis, and Liston 2003, 37) Some pay passing attention to first responders, but far

too little, and I do not recall a single one that seriously considered the role of the average citizen.

The collaboration issue is a key one; the failure of agencies to connect has been the most prominent of the many 9/11 failures. The GAO and the public interest group reports drone on about the need to cooperate, network, collaborate, or link the disparate agencies within the DHS, as well as link DHS agencies with the powerful intelligence and defense agencies outside of it. But cooperation and collaboration are not costless, and appear to be the exception in government, not the norm. The authority over the agencies that are to cooperate (i.e., the White House) calls for cooperation but does not expend the resources (attention, monitoring, political capital, clear signals) to insure it. Every agency wants the others to cooperate with it but is reluctant to cooperate with them. The fear is that autonomy is lost unless cooperation is on one's own terms, which isn't cooperation at all. The reports all fail to confront this issue. Witt was able to get many agencies to cooperate within FEMA, and a good part of this may have been his skills, but the largest part could have been the support he had from the White House. Such support can turn even an administrator of modest talents into what appears to be unusual talents. (Organizations can be neither designed nor led in a way that requires exceptional personnel.) Therefore, the reports might have noted that given the interests of individual organizations, collaboration requires determined oversight and insistence by the head of the executive branch, the president. Asking the organizations to cooperate more, as the reports do, evades placing responsibility for their failure to cooperate where it belongs.

The GAO itself issued one hundred reports on homeland security even before 9/11, and in the first three years after that issued more than two hundred additional critical reports. Every week or two it cited a string of failures (and a sprinkling of successes) of the department and, while acknowledging the insufficient funding, it criticized the DHS's poor fiscal management and waste, as did the DHS's own inspector general office. Within a year of its creation (the legislation was signed November 25, 2002) the GAO desig-

nated it as a "high risk" agency, indicating serious performance problems.

It was not only large (180,000 employees) and diverse (twenty-two agencies with 650 separate computer systems to integrate), but so many of the agencies it took in were already high-risk agencies by GAO standards—that is, highly inefficient and poorly meeting the challenges in security and nonsecurity functions as well. The moves were unlikely to increase their performance, since some nonsecurity functions (e.g., fishing rights, computer crime, tariffs) might have benefited from staying close to other agencies that were not brought in to the new department. The new department merged agencies that, along with their security roles, had responsibilities for such unrelated activities as fisheries, mapping floodplains, river floods, animal diseases, livestock inspections, a national registry for missing pets (yes!), energy reliability, computer crime, citizenship training, tariffs on imports, drug smuggling, and the reliability of telephone networks. The potpourri of unrelated activities exceeded that of any previous large government mergers.

Interorganizational cooperation or coordination is familiar in the business world and among voluntary organizations. It takes work, but there is nothing impossible about it. Large corporations cooperate routinely on political and legislative matters (which is why some theorists refer to the "power elite"). Voluntary organizations, even when competing for funds or clients, cooperate informally and through formal superorganizations that they fund and join. Networks of small firms in Northern Europe, Japan, and to some extent the United States (Silicon Valley, biotech firms) cooperate extensively and productively, even while competing. And there are radically decentralized firms such as Johnson & Johnson that take steps to insure cooperation and coordination among their hundreds of small, independent units. It is far from impossible. But these cooperative enterprises are networks; DHS is not that form of organization. It consists of agencies with diverse histories thrown together, with only parts of each agency tasked with, or assigned, roles in a common enterprise, and subject to the oversight

of several committees. In such a case, cooperation will be extremely difficult, perhaps more difficult than before the reorganization.

Public sector transformations are more difficult than those in the private sector, the head of the GAO, the comptroller general, wrote in a letter to a congressional committee, because organizations in the public sector must contend with more power centers and stakeholders, have less management flexibility, and are under greater scrutiny. Furthermore the top officials are typically political appointees who do not stay for long. (Walker 2004) Even in the private sector, where things are easier, his letter notes that more than 40 percent of executives in acquired companies leave within the first year, and 75 percent within the first three years. It even takes from five to seven years, according to research, to make mergers and acquisition work in the private sector, with all that sector's advantages.

In December 2004, Clark Kent Ervin, the inspector general of the DHS (a watchdog division within the department) was not reappointed. He was a Harvard University–trained lawyer from Texas who had worked in the first Bush administration. But he had issued many critical reports accusing DHS officials of ineptitude and of fraud, including a charge that almost $50 million in excess profits were paid to the Boeing Company. (Ross and Schwartz 2004) He repeated some of his criticisms in an op-ed piece in the *New York Times* shortly after he was fired (Ervin 2004) and gives more blow-by-blow details in his book about his time in the department. (Ervin 2006) The failure to reappoint him was an alarming development, and, as we shall soon see, the first of many instances of using the terrorism crisis to countenance favoritism and corruption.

It is not clear as yet what the reorganization meant to the agencies that were moved, except that they were expected to take on new duties or increase their security efforts. In many cases it may only have meant a change in the letterhead, and the personnel continue to use their contacts with other agencies and go about their business. One cannot imagine any great changes in, say, the Plum

Island Animal Disease Center, other than increasing their research on the deadly substances that terrorists might use. The Federal Protection Services, moved from the General Services Administration, may have received some new weapons and was told to have some meetings with the Office for Domestic Preparedness, moved from the Justice Department, but one can't imagine much new synergy from such contacts or much new energy. (An exception, pointed out me by Laura Donohue, is the Exercise Division, transferred from the Department of Justice, which runs three hundred high-quality exercises a year. Another was to bring together several offices in various agencies that deal with border control.) But to the extent that new security responsibilities are added to the agencies, as must be the case with almost all of them, one can imagine that all the agencies that are folded into the department will be having a harder time doing the jobs they were originally designed to do.

The improvements that have been made, in border security, airline security, immigration checks, and a small beginning in port security, all could have been made *without* any major reorganization, if so directed by the White House. Indeed, the former inspector general wonders, considering all the failures and the loss of an intelligence role, "why it is, exactly, that we still have a Department of Homeland Security." (Ervin 2006, 189)

### New Roles for Old Agencies

Even if the DHS had not been formed and a coordination unit in the White House were responsible for our response instead, there had to be problems with asking many agencies to add terrorism to their task list. The Immigration and Customs Enforcement service is designed to let people in to the country, particularly those whose skills are deemed to be in short supply. It is now asked to deny entry to suspected terrorists but has little expertise to do this. The Border Control agencies are designed to prevent smuggling and illegal entry but have little capability to identify terrorists. Thus new tasks must be learned, and connections be made to parts of organ-

izations they had little contact with (the FBI's counterterrorist divisions and the CIA).

One basic problem is making fruitful connections between the relevant agencies. Finding the right organization is not easy. If you suspect a threat to your chemical plant, do you go to the local police, the FBI, or the EPA—or the Coast Guard or border control if it's near the water or a border? Whichever one you pick then has the problem of deciding which of the others, and how many, should be involved. (Information should be shared, and dots connected.) If there is a dispute among them as to what actions to take, or what resources a unit should supply, what is the next level that would resolve the dispute? The organizations contacted may not accept your authority if you demand specific resources, or you may not be sure who has the resources you think necessary.

Even the question of what person you contact in the organization may be problematical. Take, for example, the frustrations of the Centers for Disease Control and Prevention (CDC). Though (fortunately) they were not transferred to the DHS, the example illustrates the interorganizational problems of combating terrorism. Shortly after 9/11, an anthrax attack occurred, sending the lethal substance through the mail, where contaminated sorting machines spread it randomly to hundreds of addresses in addition to the intended addresses. The experts who understood anthrax were at the CDC, so it was sensible to call them in, but they were caught off guard. What they knew about anthrax came almost exclusively from agricultural settings, where it occurs naturally and infects animals and farm workers. "Everything we knew about the disease just did not fit with what was going on. We were totally baffled," said an official of the public health system. (Altman and Kolata 2002) Anthrax had always been an agricultural problem, but now it was a criminal one, thus requiring contact with the FBI. But, an official at the CDC told me, there was no reason for them to have any established links with the FBI. The FBI dealt with criminals, the CDC with microbes and victims. Still, one might think that networking should be easy; just get on the phone and call the FBI. ("Listen to the options carefully; they have changed.") But who

would they call? Since the FBI had no experience with what looked like a criminal epidemic, its help lines were of little use. ("Your call is important to us; please try again.")

This official had, by chance, struck up a friendship with an FBI agent as a result of a totally unrelated conference a year before. She had the agent's number; she called and explained the problem, and he was able to run interference for her up the FBI hierarchy to the appropriate level and office. (It is unclear how the Centers for Disease Control escaped being folded into the new DHS; given the threat of bioterrorism, they would seem to belong there more than, say, the Plum Island Animal Disease Center. Making them a part of the DHS would not have made it any easier for them to contact the FBI, even if the FBI had been also made a part of the department. Given their unique character and extraordinary importance and expertise, one can be grateful that the CDC remain reasonably independent.)

Since the attack was unpredicted, there were no established routines to call into play. Not only had the CDC seldom dealt with the FBI, they had had no reason to deal with the post office, state and local police agencies, and so on. And not only were there layers of phone numbers that needed to be searched, but the officials dealing with the attack had no authority over other agencies such as local post offices or police to get them to do what they thought needed doing once they found them. Agencies all have their organizational interests, jurisdictions, and clout. Networks are not hierarchical; the units have autonomy. An official of the CDC said they were unprepared "for layers and levels of collaboration among a vast array of government agencies and professional organizations that would be required to be efficient and successful in the anthrax outbreak." (Altman and Kolata 2002)

Even when you successfully network, what authority do you have? A Connecticut state health officer, where there was a mysterious case of anthrax illness, said "we were very much aware that we had no jurisdiction over federal facilities whether it was the V.A. or the post office." (Altman and Kolata 2002) Merging twenty-two agencies will not solve problems such as these, nor will

injunctions to establish "clear lines of authority" or to "cooperate." As the ample disaster and emergency response literature shows, coordination and cooperation requires (after the central authority lays out a meaningful mission and exercises oversight) frequent drills, exercises, simulations, and meetings where diverse agencies get a chance to see each other's point of view, establish personal contact, and build trust. This is new work, and in turn requires increased budgets. Even the efficient agency is not likely to have slack funds to handle a new unfunded mandate. Yet, as noted above, the first-responder agencies are grossly underfunded. Creating a DHS may have made it harder to direct the new terrorism funds to official first responders—larger interests got there first—and appears to have even diverted FEMA funding from them.

### Organizational Problems: Displacement of Missions

A big chunk of the new Department of Homeland Security contains an agency we discussed in the last chapter: the Federal Emergency Management Agency. What would happen to its traditional concern, natural disasters? The fate of programs concerned with natural disasters under DHS was a concern from the beginning. FEMA director Joe Allbaugh was asked early in 2002 about this and told Congress that the traditional role of FEMA would not be affected. But even Republican lawmakers were not convinced. Representative Don Young (R-AK), chairman of the House Transportation and Infrastructure Committee, said that if the homeland security secretary wanted to redirect the agency and focus on preventing terrorist attacks, he could reduce "other [FEMA] missions and direct those resources entirely to security." Congressman Young had good reason to think this possible: This is what the first director that President Reagan appointed did, as we have seen. To forestall this, the chair of the Select Committee on Homeland Security, Richard Armey (R-TX) redrafted the White House proposal to keep FEMA primarily an agency dealing with natural disasters. The White House objected, suggesting that a displacement of its

mandate could indeed have been in the cards. Some senators and the highly regarded former FEMA head, James Witt, were opposed to putting FEMA in the new department, but it was moved. (Pincus 2002) Nevertheless, when Florida was hit by a succession of four major hurricanes in 2004, there was no outrage about FEMA's performance. As we shall see, there were strong political reasons to pay attention to those hurricanes and be prepared. But otherwise, the agency had been weakened.

### Organizational Challenges:
### New Tasks, Few New Resources

A further problem was what happened to the agencies when they were transferred to DHS, and given added homeland security tasks, but expected to continue with their usual ones, often with no significant budget increase, if any. In Pittsburgh, where the Coast Guard helps control traffic on the busy Ohio River, homeland security activities had accounted for 10 percent of Coast Guard activity; it now grew to 50 percent. This meant cutting other activities, such as assisting boaters and acting as traffic cops on the crowded river. The Coast Guard's effort in drug interdiction declined by 60 percent after 9/11, and time invested in preventing an encroachment on American fishing territories and enforcing fishing rules shrank 38 percent. (Kettl 2004, 39–40) One might say that at least the Coast Guard changed its priorities, but increased funding for security could have left the old activities in place if they were deemed necessary.

The U.S. Border Patrol is an example of inadequate funding coupled with new tasks. It hasn't had much to do on the 4,000-mile border with Canada. But a terrorist was caught (almost accidentally) bringing explosives over the border to be used in an attack on the Los Angeles airport. Then it was learned that some terrorists used Canada as a port of entry (though not any of the 9/11 terrorists). The Border Patrol, in 2002, had only 330 agents supported by one analyst to intercept illegal crossings of the 4,000

mile border. In the preceding twenty years two hundred had been cut in government downsizing efforts. Half the inspection booths were simply closed. (Wise 2002a) As of April 2004, miles and miles of the border consisted of dense, overgrown brush where before there had been cleared spaces. In 2004, it finally received some more resources for an impossible task.

A serious problem has emerged that concerns the critical area of official first responders—police, fire, and emergency medical personnel and various voluntary and homeowners associations. As mentioned previously, the title of a 2003 Council of Foreign Relations task force report summed up the problems: "Emergency Responders: Drastically Underfunded, Dangerously Unprepared." (Rudman, Clarke, and Metzl 2003) The task force declared the underfunding by government at all levels to be extensive and estimated that combined federal, state, and local expenditures would have to be tripled over the next five years to address this unmet need. Covering this funding shortfall using federal funds alone would require a fivefold increase from the current level of $5.4 billion per year to an annual federal expenditure of $25.1 billion. Nor would these funds provide gold-plated responses; they would go to essentials. For example, the executive summary gave these examples of deficiencies:

- On average, fire departments across the country have only enough radios to equip half the firefighters on a shift, and breathing apparatuses for only one-third. Only 10 percent of fire departments in the United States have the personnel and equipment to respond to a building collapse.
- Police departments in cities across the country do not have the protective gear to safely secure a site following an attack with weapons of mass destruction.
- Public health laboratories in most states still lack basic equipment and expertise to adequately respond to a chemical or biological attack, and 75 percent of state labs report being overwhelmed by too many testing requests.
- Most cities do not have the necessary equipment to deter-

mine what kind of hazardous materials emergency responders may be facing. A study found that only 11 percent of
fire departments were prepared to deal with the collapse of
buildings with more than fifty inhabitants, 13 percent were
prepared for chemical or biological attacks, and only 25
percent had equipment to communicate with state or federal emergency-response agencies. (Shenon 2003a)

Furthermore, the funds that the federal government did allocate
for emergency responders were sidetracked and stalled due to a
politicized appropriations process, the slow distribution of funds
by federal agencies, and bureaucratic red tape at all levels of government, according to GAO reports.

Congress has played a substantial role itself in hampering the effort. The $3.5 billion promised by the White House in January
2002 for first responders in the state and local governments fell victim to partisan squabbles in Congress, not being approved until
more than a year later, in February 2003. Given congressional control over the DHS, this was to be expected.

A glaring example of thoughtless, on-the-fly reorganization occasioned by the birth of the new department concerned a particularly important sector of first responders—the effective National
Disaster Medical System (NDMS). It deployed and coordinated
volunteer teams of doctors, nurses, and other medical personnel in
a crisis, some seven thousand volunteers. It had been moved from
the Department of Health and Human Services to FEMA by the
original DHS designers. HHS felt it belonged in HHS, not in
FEMA, and tried to wrest control of the NDMS during Hurricane
Isabel (September 2003). It did not do well in FEMA and was
starved of resources. Its paid staff had shriveled from 144 to 57
and did not even include a physician. NDMS volunteers complained about FEMA's unpaid bills, faulty equipment, and intran-
sigent leadership. The National Association of NDMS Response
Teams sent a harsh letter to Ridge's successor, Michael Chertoff. It
warned that two years after their move to FEMA, they were less
prepared than ever: "We feel that the identity of the NDMS is

being lost via FEMA's efforts to 'swallow' NDMS functions, rather than support them. . . . During transition, it has been fragmented, reduced, and relegated to a position without the authority, staff, resources . . . or systems in place at FEMA to move forward with the most fundamental of readiness and critical mobilization issues." (Grunwald and Glasser 2005) This is one of many instances wherein the DHS has made us less safe.

### Organizational Uses

Organizations, as I have often noted, are tools that can be used by those within and without them for purposes that have little to do with their announced goals. A new organization such as the DHS invites use. As soon as the department was established, the corporate lobbying began. Four of Secretary Tom Ridge's senior deputies in his initial position as Assistant for Homeland Security at the White House left for the private sector and began work as homeland security lobbyists, as did his legislative affairs director in the White House. The number of lobbyists that registered and listed "homeland," "security," or "terror" on their forms was already sizable at the beginning of 2002, numbering 157, but jumped to 569 as of April 2003. One lawyer for a prominent Washington DC law firm was up-front about corporate interests. He mentions in his online résumé that he authored a newsletter article titled "Opportunity and Risk: Securing Your Piece of the Homeland Security Pie." (Shenon 2003b) It is a very large pie indeed.

A Web document, "Market Opportunities in Homeland Security," introduces buyers to the "$100 billion" homeland security marketplace, for $500.00 plus shipping. Less exuberant in its predictions, a Frost & Sullivan report indicates the industry generated $7.49 billion just in 2002, with total market revenues of $16 billion estimated for 2009. Frost & Sullivan is an "international growth consultancy," found at www.frost.com. A report from Govexec.com by Shane Harris, titled "The Homeland Security Market Boom," issued less than six months after 9/11, documents

the aggressiveness of U.S. business in flocking to the new funding source. "Every good company out there can take what they do and reposition it for homeland defense," says Roger Baker, the former chief information officer of the Commerce Department, who is now with a private company. (Harris 2002) In 2006, less than three months after registering as a lobbyist, former attorney general John Ashcroft was developing a practice centered on firms that want to capitalize on a government's demand for homeland security technology, and had banked at least $269,000 from just four clients. Observers said it was virtually unprecedented for a former U.S. attorney general to start a lobbying firm. (Staff 2006a) An article in *Rolling Stone* lists a number of products that lobbyists procured for their client: Tiptonville, Tennessee (population 7,900) received $183,000 for an all-terrain vehicle, defibrillators, and hazmat suits for the volunteer fire department. The mayor explained: "If I were Al Qaeda and wanted to get to Memphis . . . I'd come to Tiptonville. No one would expect me there." Mason County, Washington (population 54,000), got $63,000 for a biochemical decontamination unit that no one has been trained to use. Bennington, New Hampshire (population 1,450), got nearly $2,000 for five chemical weapons suits. The police chief said he saw no specific threats but it was being offered so he figured that he would get on the bandwagon. (Klinenberg and Frank 2005) Converse, Texas, used its money to transport lawnmowers to the annual lawnmower race, and the mayor of Washington DC, Anthony Williams, certainly a city that is a legitimate terrorist target in contrast to these others, spent $100,000 to produce a rap song on emergency preparedness and $300,000 on a computerized cartowing service. (Ervin 2006, 186)

As bad as all this was, what Clark Ervin's chapter on fraud and waste within the DHS, principally by the Transportation Security Administration (TSA) and the bureau of Immigration and Customs Enforcement (ICE), reveals is over-the-top. Contractors big and small, some of them national figures such as Boeing or Unisys Corporation swarmed over the agency, charging extravagant fees with cost-plus contracts that had no records of expenditures. The

failures of DHS managers seem almost willful, and in some cases involved outright fraud. Ervin's office, according to his account, was stymied by DHS top management in his attempt to establish accountability. (Ervin 2006, Chap. 9) It continued, of course, after he was not reappointed, with the TSA literally losing track of millions of dollars every month in 2005. (Phillips 2005)

There were other within-government uses, too. Our chapter on natural disasters noted that presidential declarations of disaster areas, and the federal funds that followed, varied directly with the political importance of the area to the president of the time. Shortly after 9/11, Congress passed the USA Patriot Act. Senator Patrick Leahy (D-VT), then chairman of the Senate Judiciary Committee, wrote in the criteria for distributing some $13.1 billion among the states. His committee used a formula long in use for distributing much smaller funds, one that favored the small states. The small states now resisted any change in the formula, and could do so since they had the power in the Senate. The funding was almost exactly in reverse order of the threat. (The degree of threat has been assessed by a nongovernmental research organization using sophisticated probability models.) The ten highest amounts went to states and districts with the least threat, except for Washington DC, where the congresspeople work. Thus Wyoming received $61 per person, but California only $14. Alaska, hardly a target for terrorism, received $58 per person, while New York, the target of six separate plots by Islamic terrorists in the last decade, got only $25. (Ripley 2004)

This point deserves elaboration. The formula meant that 40 percent of the funds had to be divided equally among the states, regardless of population. The executive branch had discretion over the remaining 60 percent, and it did at least somewhat better by distributing it according to each state's population; but it too did not distribute the funds according to the risk the state's population was exposed to. In early 2003, Congress announced a plan that might rectify the situation, a new $100 million grant for "high threat" urban areas only. New York City, for example, would get 25 percent of it. Immediately, Congress pressed the administration

to increase the size of the lucrative handout and also increased the number of cities at risk. Disasters are funding opportunities. Soon fifty cities, perhaps politically important to the administration, were designated as "high threat," and while the size of the grant grew to $675 million, New York City received only 7 percent instead of 25 percent. Its funds were doubled, but many more low-risk cities were now funded. Democrats charged that the Bush administration allowed this to happen because it doesn't have a constituency in the big cities. (Ripley 2004, 37)

Despite charges that the basic formula in use distorted federal outlays and was only partially rectified by the grants to specific cities, the formula was still in place at the end of 2004. The House approved a bill to have the funding formula reflect the risks the states faced, and the White House, to its credit, made a similar request in its 2005 budget. But the Senate would have none of it. Senator Leahy is a member of the powerful Homeland Security Appropriations Subcommittee, and his state of Vermont gets a handsome $54 per capita. He curtly reminded Secretary Ridge of the power of small states such as his. "I have to say, I was really disappointed that the President's proposed budget . . . drops the all-state minimum formula," he said. "That would affect all but, I think, one or two in this subcommittee." He charged that the administration "wants to shortchange rural states." The funding bill, according to Senate aides, "would go nowhere," and it didn't. (Ripley 2004)

Wyoming is the number one recipient of homeland security money per capita. Asked about this, the typical remark was that "our citizens deserve the same kind of protection that they are afforded in other places in the country." This was from the chief of police in Douglas (population 5,238), who had just received a new $50,000 silver RV that serves as an emergency operation command center, paid for with federal dollars. Firefighters in Casper, Wyoming, even denied they were less at risk than, say, New York City residents. "No one can say Casper can't be a terrorist target." Wyoming had the largest budget surplus, as a percentage of budget, of any state in the nation. Yet the seriously in-debt state of California

spends five times as much, per person, of its own money on home-land security—taxes its citizens pay—as does Wyoming. (Ripley 2004, 37) The unrepresentative character of the Senate (Dahl 2003) and the parochial interests of the citizens of small states, who expect their senators to bring in the federal dollars, will make it difficult to respond to our vulnerabilities.

Finally, in December 2004, the DHS was able to get around Senator Leahy's bill, and announced a new formula that focused on cities rather than states, over the protests of small states, that went partway to matching funds to threats. New York City was the biggest winner, going from a $47 million grant to $208 million, and Washington DC, Los Angeles, Chicago, and Boston got smaller increases. (Lipton 2004) Senator Diane Feinstein (D-CA) tried again to further increase the proportion of funds to risky areas in 2005, but Senators Collins (R-ME) and Lieberman (D-CT) blocked it. As of this writing, the issue is at least still alive.

### Another Organizational Use: De-unionizing

The new department offered opportunities to further presidential agendas unrelated to the terrorist threat. Though President Bush did not favor the department and its massive movement of twenty-two agencies, it provided an opportunity for what appeared to be an attack on civil service. President Bush immediately demanded that Congress strip all employees who would be transferred of civil service status. Liberals and labor saw this as an attack on the eighteen different government unions, and that it would reduce the amount of union membership in the government significantly. The president argued that because of the unique, nonroutine nature of defense of the homeland, the department needed to be free of civil service and union restrictions on terminating employees. The matter dragged on until March 2003, when the last day for final comment on the proposal arrived. A ninety-one-page comment from three powerful unions representing about one-quarter of the department's workers arrived. The unions had squared off for a fight.

The DHS and the Office of Personnel Management proposed regulations that would cover 110,000 of the department's 180,000 employees, affecting how they would be paid, promoted, and disciplined. It would become a model for revamping civil service rules in the rest of the federal government. Pay would be linked to performance (political performance and less aggressive bargaining, the union argued), union bargaining rights in several areas would be restricted (e.g., deployment of workers and use of technology), and the government would speed up and tighten the disciplinary process. (Barr 2004) There was a lengthy standoff, and as of October 2005, the issue is still in dispute since a federal judge, for the second time, ruled the new personnel rules invalid. (Staff 2005e) Disasters are opportunities.

### Departure of Key Personnel

The departure of seasoned terrorist experts started almost immediately. Rand Beers had thirty-five years of experience in intelligence; he had replaced Oliver North, who was the director for counter terrorism and counternarcotics in the Reagan administration. He spent seven months in the new department, and five days after the Iraq invasion in March 2003, he resigned. Three months later, he told a *Washington Post* reporter of his disaffection with the counterterrorism effort, which was making the country less secure, he said. The focus on Iraq, he said, "has robbed domestic security of manpower, brainpower and money." (Blumenfeld 2003) Agreeing with many counterterrorism experts, he saw the minimalist Afghanistan war as only dispersing Al Qaeda and not pursuing it enough to disable it, and the maximalist Iraq war as recruiting terrorists. Another disaffected expert, Richard Clarke, left the NSC in February 2003, just before the Iraq invasion, saying the same thing. His revelations about the misdirected, underfunded, and bureaucratically incompetent response to the terrorist threat, *Against All Enemies,* made the bestseller lists in April 2004.

Other experts departed or would not be recruited. A *New York*

*Times* story in September 2003, six months after the start of the department, reported two top officials leaving. "So few people want to work at the department that more than 15 people declined requests to apply for the top post in its intelligence unit—and many others turned down offers to run several other key offices, government officials said." (Mintz 2003) The administration announced that 795 people in the FBI's cyber-security office would be transferred to the DHS, but most decided to stay with the more reliably funded, higher-status FBI, and only twenty-two joined the new department. (Mintz 2003) Flynt Leverett, who served on the White House National Security Council for about a year until March 2003 and is now a fellow at the Brookings Institution, observed, "If you take the (White House) counterterrorism and Middle East offices, you've got about a dozen people . . . who came to this administration wanting to work on these important issues and left after a year or often less because they just don't think that this administration is dealing seriously with the issues that matter." (Drees 2004) In a union survey of eighty-four union personnel in the DHS, 80 percent said it was a "poorer agency," and 60 percent said they would leave if they could get the same salary in another agency; and the GAO rated morale at the DHS as one of the lowest of any government agency. (Elliston 2004) For other examples of departures, see *Against All Enemies*. (Clarke 2004) The "brain drain" continued into 2006 and appeared to be increasing. (Hall 2006; Lipton 2006f)

### Centralizing to Combat a Decentralized Enemy

A final concern with the new department was the emphasis on centralization. Unfortunately, the immediate response of most politicians and even some administrators to signs of poor coordination, indifference to changing environments, and new tasks is to rein everyone in, centralize, and give specific tactical orders. I expect this is what Senator Lieberman and others had in mind when they wrote their proposed bill, which was similar in this respect to the

one the president's aides later drafted. It was easy to react this way to incredible stories of bungling and mismanagement. Bringing twenty-two agencies under one command seemed quite sensible, since they rarely had worked together. In the intelligence reorganization legislation, which we will come to later, creating an intelligence czar was the easiest response to the credibility-challenged CIA and the balkanized fiefdoms of the FBI.

In contrast, some commentators recommended that the structure of the new department and of the intelligence agencies should match the structure of the enemy. I do not know of any compelling theoretical arguments to support this view, but it makes considerable sense nevertheless and is worth exploring. The threat comes from highly decentralized terrorist networks, while the response comes from two newly centralized, hierarchical agencies. Current Islamic terrorism of the Al Qaeda type involves unpredictable acts by cells in loosely coordinated networks. To defeat the networks requires on-the-ground operatives with maximum autonomy that can infiltrate the networks and also exercise close surveillance, striking only when an operation is imminent. A centralized response is to bomb any suspected targets, raid them, and round up anyone who looks suspect. To have the best response to domestic threats is to allow considerable autonomy to border control agencies, to airlines and other transportation agencies, and, when the attack comes, to official first responders. (For devastating critiques of the 9/11 Commission's view of the role of citizens and official first responders and its recommendation to centralize what should be decentralized, see K. Tierney 2005.)

Since intelligence was to remain outside the DHS, I would recommend a coordinating role for the head of intelligence, instead of a czar, whose responsibilities would be to collate information from the intelligence community; conduct frequent reviews, surveys, and evaluations of the separate agencies; and recommend budgetary changes to the White House. The intelligence coordinator should not have distracting operational duties, such as George Tenet did as also directing the CIA. Intelligence would remain decentralized. The head office of the DHS would behave similarly—

coordinating, reviewing, evaluating, and handling the budgets of the diverse agencies. Indeed, most of the twenty-two agencies would not need to have been moved. With strong White House support for the head of intelligence and the head of a homeland security coordinating office, a decentralized response need not mean chaos or balkanization, but agency empowerment.

## FEMA UNDER A BUSH

We will leave the DHS as a whole at this point, and turn to FEMA, a cabinet-level small agency that had been swallowed up by the DHS whale. Though FEMA appears to have done well in the succession of hurricanes in Florida in 2004, two things made this response unusual. Florida has yearly hurricanes, and Governor Jeb Bush had a well-organized state response team that was able to assert independence from FEMA when necessary. Second, it was also a politically sensitive state with a Republican governor, offering great opportunities for spreading federal grants, even over areas that were not touched by the hurricane. This was a crucial election year for President Bush, and Michael Brown, the head of FEMA, did his job well. In a six-week period in August and September 2004, emergency-supply trucks were pre-positioned to deliver ice, water, cots, blankets, baby food, and building supplies. The magazine *Government Executive* reported: "Seldom has any Federal agency had the opportunity to so directly and uniquely alter the course of a presidential election. . . . Seldom has any agency delivered for a president as FEMA did in Florida this fall." (Klinenberg and Frank 2005)

Katrina was different. It was enormous, the state was not politically important, there was little pre-positioning, and FEMA had had another year in which to decay. Michael Brown, appointed head of FEMA in 2002, became the target of the news channels, late-night talk shows, and Comedy Central. During Katrina he managed to make a fool of himself with e-mails about what time he needed to have dinner and his success at shopping at Nordstrom

while New Orleans drowned. But *Washington Post* writers Michael Grunwald and Susan Glasser (2005) offer a far more complex view of this contradictory personality and his agency: Brown was dedicated to FEMA and to keeping it independent (after all, his career was riding upon it), and he alienated the White House and other DHS agencies and its leadership with his harsh analysis of their failings. Just as Katrina was an outsized storm, Brown and his agency are outsized failures, going far beyond the normal inevitable failures of organizations that are the subject of so much of this book. As flawed as the response to 9/11 was, all the responsible government agencies (except the Coast Guard) appeared to perform considerably worse in Katrina. This failure requires exploration and interpretation. Though unusual in their magnitude and tragic consequences, the failures of the DHS and FEMA and its director are clearly within the realm of the possible, thus strengthening the theme of this book that prevention and mitigation will always fall short, sometimes alarmingly so, and we should begin to reduce the size of our vulnerable targets.

Joseph Allbaugh was once a third of President Bush's "Iron Triangle," along with Karl Rove and Karen Hughes, but was "exiled," Grunwald and Glasser say, to head up FEMA after the 2000 campaign. Then in 2002 Bush announced that FEMA would lose its cabinet-level status and be placed under the new DHS, so Allbaugh quit. He arranged to have his deputy, Michael Brown, succeed him. Initial press reports in *Time* and *The New Republic* indicated Brown did not have a background in emergency management and had exaggerated his teaching experience in his résumé, but these were successfully contradicted by affidavits he assembled for testimony before a House select bipartisan committee investigating the hurricane months later. (U.S. Congress 2005) Brown had gathered experience in bureaucratic politics in his two years working for Allbaugh. He was determined to keep FEMA as independent from DHS as possible, but the first head of DHS, Tom Ridge, wanted it to be integrated into DHS and to be a key player in DHS, and spoke of putting FEMA "on steroids." To do so, however, meant that its focus would have to shift from an emphasis on

natural disasters to one on terrorism. Ridge stripped FEMA of its control over the millions of dollars worth of preparedness grants concerned with natural disasters.

FEMA's Project Impact was a model mitigation program created by the Clinton administration; it moved people out of dangerous areas and retrofitted structures. (Elliston 2004) For example, when the Nisqually earthquake struck the Puget Sound area in 2001, homes and schools that had been retrofitted for earthquakes with FEMA funds were protected from high-impact structural hazards. The day of that quake was also the day that the new president, George W. Bush chose to announce that Project Impact would be discontinued. (Holdeman 2005) Funds for mitigation were cut in half, and those for Louisiana were rejected. Three out of every four grants for mitigation are now spent on counterterrorism. (Much of the money spent on counterterrorism goes to corporations and private businesses; natural disaster money is more likely to be spent on training first responders, hardly a corporate feeding place.) This probably was a major blow to states such as Louisiana that are prone to weather disasters.

But the states and counties themselves may have weakened their disaster programs. Political scientist Patrick Roberts details the extensive cuts in federal funds for natural disasters, but he also makes another telling point. "State and local emergency management agencies reorganized to meet the terrorism threat to a much greater degree than has FEMA. . . . Many state emergency management agencies may have simply been too small and weak to withstand the funding and attention shift toward the terrorist threat. These agencies depend on federal and state grants for their operational budgets, and when grant criteria emphasized the terrorist threat, state and local agencies shifted their priorities. In addition, the law enforcement culture, which is more concerned about terrorism than is the natural hazards culture, is stronger in some state and local agencies than at FEMA." (Roberts 2005, 443) This may help account for some of the alarming failures at the state and local level.

Ridge also seized the Office for Domestic Preparedness (ODP) from the Justice Department. This was a major source of funding,

so it was politically valuable. The OPD did not go to FEMA, however, but to his own office in DHS, and FEMA's preparedness grants went there too. Brown objected. He sensibly noted that at the state and local level, the people responsible for preparing for disasters were the same who responded to them; it did not make sense to pry them apart. Brown appealed all the way to the White House but was overruled. (Grunwald and Glasser 2005)

FEMA also lost its grant program for fire departments, its terrorism training program, and still other grant programs. Ridge's office got the job of creating a "National Preparedness Goal" that would create likely scenarios, something FEMA had expected to do. Finally, Ridge's office got the granddaddy of them all, the National Response Plan. By the time Katrina hit, the plan had not been exercised, nor had its all-important appendix dealing with first responders been drafted.

In this environment we would not expect an appropriate response to the hurricane, but the details of the response are still alarming. For example, FEMA director Brown said on a Thursday evening TV appearance, three days after Katrina struck, that he had just learned of the plight of thousands stranded at the convention center in New Orleans without food or water. They had been there—and shown on national TV news—since Monday, but Brown told an incredulous TV interviewer, Paula Zahn, that Thursday, "Paula, the federal government did not even know about the convention center people until today." (Lipton and Shane 2005)

FEMA and Brown also did not know where the ice was. It was not pre-positioned, as it had been in Florida. Ninety-one thousand tons of ice cubes—intended to cool food, medicine, and victims in over–100 degree heat—were hauled across the nation, even to Maine, by four thousand trucks, costing the taxpayers more than $100 million. Most of it was never delivered. In an age of sophisticated tracking (FedEx, DHL, WalMart, etc.) FEMA's system broke down. Asked about the vital ice, Brown invoked privatization, and told a House panel, "I don't think that's a federal government responsibility to provide ice to keep my hamburger meat

in my freezer or refrigerator fresh." (Shane and Lipton 2005) The ice was not needed for his refrigerator but to keep the food, drugs, and medicine for the victims fresh, to treat people with heat exhaustion, and to keep the sick, old, and frail cool. There was plenty of it for Florida the year before. The drive to privatize was signaled earlier by Brown, before he was made FEMA head. At a conference in 2001 he said: "The general idea—that the business of government is not to provide services, but to make sure that they are provided—seems self-evident to me." (Elliston 2004)

For days following Katrina empty air-conditioned trucks with no supplies drove aimlessly past "refugees" who had no water or food or protection from the sun. Reporters came and went, but food and water and medical supplies did not. (Staff 2005d) The Red Cross was not allowed to deliver goods because that might discourage evacuation. (American Red Cross 2005) Evacuation by air was slowed to a crawl because FEMA said that post 9/11 security procedures required finding more than fifty federal air marshals to ride the airplanes and finding security screeners. This search was prolonged. At the airport's gates, inadequate electric power for the detectors prevented boarding until officials relented, but they still required time-consuming hand searches of desperate and exhausted people. (Block and Gold 2005) The only food—emergency rations in metal cans—was confiscated because the cans might contain explosives. (Bradshaw and Slonsky 2005) Volunteer physicians watched helplessly; FEMA did not allow them to help because they had not been licensed in the state. (J. Tierney 2005) Without functioning fax machines to send the required request forms, FEMA would not send help that local officials begged for. (Apparently no one at FEMA dared violate the rules, even in such an emergency.) Perhaps a fifth of the New Orleans police force simply quit—exhausted and discouraged, their own homes gone—or were themselves looting. A large National Guard force hid behind locked doors in the convention center, saying "there were too many of them for us to help" and when they went forth on their mission they sneaked out to avoid the hungry evacuees, saying they needed their food for themselves. (Haygood and Tyson

2005) A Navy ship with transport and hospital facilities idled off-shore, waiting for days to be called. Almost five days after the next hurricane, Rita, struck, at least one severely damaged Texas town remained without any outside help—out of power, water, and food—with an alerted TV camera crew being the first to arrive. (Rita came ashore just a few days after Katrina.) For other tales of similar failures, see "Hurricane Katrina as a Bureaucratic Nightmare," by Vicki Bier. (2006) As noted in chapter 2, the failures were not only official ones. Because Murphy Oil failed to prepare its tanks properly for the storm, 1,800 homes were made uninhabitable and the St. Bernard Parish of New Orleans was polluted by 1.1 million gallons of oil.

A particularly disturbing report about the new DHS came out in the congressional investigations of the hurricane response. The department had set up a Homeland Security Operations Center that costs about $56 million a year to run and provides up-to-date news on impending disasters for more than thirty agencies. With Brigadier General Matthew E. Broderick, a veteran of Vietnam and Somalia, in charge, the center watched Hurricane Katrina bear down on the Gulf Coast on Monday, August 29. The dispatches begin rolling in on Monday morning, reporting major flooding in some parts of the city, people calling for rescue from rooftops and attics, ten feet of water already in some areas, and the flooding increasing. Warnings of breached levees came in from the Coast Guard, the Army Corps of Engineers, and the Transportation Security Administration during the day. But before he left to go home to bed about eleven o'clock that evening, Broderick saw a television report showing that there was drinking and partying in the French Quarter, which was on high ground and never actually flooded. That report apparently reassured him. He told investigators that he was not surprised to hear of flooding during a hurricane, that was expected, but that the operations center's job was to "distill and confirm reports." "We should not help spread rumors or innuendo, nor should we rely on speculation or hype, and we should not react to initial or unconfirmed reports which are almost invariably lacking or incomplete." So he did not place a call Mon-

day evening to Mr. Chertoff at the DHS or to the White House. It is probably irrelevant that the head of FEMA unilaterally decided not to work with the center and didn't tell Broderick directly what FEMA had witnessed that day, since Brown himself was ignoring television coverage. (Lipton 2006a) The Department of Homeland Security Operations Center is a poster child of the organization, used to advertise the country's preparedness.

Katrina also triggered the biggest deployment in the National Disaster Medical System's history. One official called the result "a national embarrassment." "In an after-action report, a NDMS team leader Timothy Crowley, a doctor on the Harvard Medical School faculty, called the deployment a 'TOTAL FAILURE.' Crowley's team was summoned late, then sent to Texas instead of Louisiana, then parked in Baton Rouge for a week while New Orleans suffered." The team was finally sent to the disaster zone and was immediately overwhelmed by the demands for help, but no additional help was available. It later found out that a host of other teams "had been sitting on their butts for days waiting and asking for missions" said Crowley. (Grunwald and Glasser 2005)

### Accounting for the Katrina Response

FEMA was not the only organization to fail so massively in Katrina and Rita, but it, and its parent organization, the Department of Homeland Security under Michael Chertoff, was certainly the key one. Can we attribute the failure to the evisceration of FEMA under the Bush administration? Did its enfeeblement also enfeeble the response of the National Guard units, the military when it was called in, and local and state agencies? Or was it the size of the hurricane? We have some data points for comparison. The response to Rita, arriving a few days after Katrina, has been declared much better by some news stories (Hsu and Hendrix 2005; Block and Gold 2005), and almost as bad by others. (Staff 2005f) Rita should have been easier. It was less destructive; citizens were more likely to evacuate early based on the experience with Katrina; no major cities were hit;

top FEMA officials would be unlikely to again be unable to get the president's attention; and state guards and the military were already mobilized. The failures in Rita were not encouraging.

Another data point is the response of FEMA to the four hurricanes that hit Florida in 2004. The only really critical news stories about the response refer to the large amounts of money distributed to areas that suffered no hurricane damage, widely attributed to Republicans' interest in keeping that state in its column. FEMA approved payments in excess of $31 million to Florida residents who were unaffected by the 2004 hurricanes, for example. (Leopold 2005) (Staff 2005a) (As noted in chapter 3, political scientists have found that nearly half of all disaster relief is delivered on a political basis rather than by need.) But FEMA was actually blocked from playing more than a role as a resource provider in the state, on the order of Governor Jeb Bush. Florida had more experience with hurricanes than Louisiana, was wealthy enough, and was well-connected enough to fund extensive programs and exercises to cope with familiar disasters; and while four hurricanes in a row was unprecedented, none were of the magnitude of Katrina. As noted, the hurricanes occurred just before the critical national presidential election, and FEMA made sure that it was ready. One possible explanation for the failure is the indifference of the White House as the accurately forecasted hurricane headed toward New Orleans and swept through the three states. The president was on a golfing vacation, was briefed on the danger in a conference call, but assumed all would be taken care of in the biggest threat to a large city in recent memory. None of the cabinet members changed their routines. Michael Chertoff, the head of the DHS, saw no need to declare a federal emergency until days after the storm hit; and the head of FEMA said his staff did not inform him about the deplorable conditions in the convention center, even though some of them must have been watching, along with the rest of the nation, the endless TV news coverage. It is hard to explain this indifference, especially in what is said to be a very politically astute administration that could capitalize on a disaster with a vigorous response.

One explanation could simply be the disarray in the DHS in general—a department the administration did not want, but when forced on it, used it to reward political loyalists, privatize government, defeat civil service, and give out contracts to business friends. But there was more than disarray; the uses to which it was put shaped its response. The DHS had been fabricated in haste and neglected after its launch. But it also reflected deeper political values. As noted earlier, Brown shared the administration's view that less government was better than more. It was "self-evident" to Brown that the government was not to provide services but only to see that the private sector provided them. FEMA would not be proactive in the face of the predicted disaster—an assertive role for government interference in the private sector—but would wait until help was properly requested (except in politically sensitive Florida). With that view of the mission of FEMA at the top, it is hardly so surprising that urgency was lacking at the bottom.

There was also more than an expression of conservative political values at work. One of the most extraordinary things about this disaster is the extent to which people at the lower levels of the responding organizations simply did not use common sense. Neither panic nor being "overwhelmed" will answer the following questions: Why did they insist upon full documentation when this was not possible because the fax machines were not working? Why did they force qualified medical personnel to stand aside because they could not legally engage in medical care in Louisiana? Why did they insist that air marshals had to be found to fly on the airplanes that carried out desperate evacuees? Why did they insist on time-consuming searches because the electronic monitors had no power? Why did they remove tins of emergency food from the evacuees because of regulations that were irrelevant to this emergency? Why did they not allow empty buses to pick up refugees? Why did they not require that National Guard trucks idly driving by thirsty evacuees carry water with them and distribute it? And the list goes on and on.

These are not instances that can be explained by the overwhelming nature of the disaster; the officials could have behaved

differently, and the force of the storm or the destruction is irrelevant. They also go well beyond the issue of "prosaic failures" that all organizations are subject to. (Clarke and Perrow 1996) These behaviors do not involve panic, enormous overload, or unfamiliar tasks or settings, conditions that usually account for failures in unprecedented events. They involved going by the rules. Rather than being flexible and innovative, which is possible even when the challenge is overwhelming, these personnel appeared to revert to rote training, insistence on following inappropriate rules, and an unusual fear of acting without official permission. This is what needs explanation.

I would suggest that as the top ranks of the agency lost experienced personnel with high morale and commitment, and were replaced by political appointments with no professional experience in emergency management, the next level would gradually lose confidence in their superiors, and their morale would slacken. I know of no statistics regarding FEMA, but nationally the Bush administration had increased the number of political appointees for government agencies by 15 percent between 2000 and 2004. (Writers 2005a) (In President Clinton's second term, the percentage of political appointments declined.) FEMA has always had political appointees; most agencies do and some political appointees may even have experience in their field. But it was widely believed that the number of inexperienced political appointees rose dramatically during the Bush administration. Even if the increase in FEMA was only the average 15 percent, it would have an impact. The problem continues as of April 2006. Looking for a replacement to head FEMA, several people with extensive experience were recruited but all of them turned the Department of Homeland Security down. Most of them cited the failure of the White House to establish clear goals and a clear role for FEMA within the DHS, including ambiguity about the importance of natural disasters as compared to terrorist threats. (Lipton 2006f)

In time, the low morale of upper managers who were not political appointments would spread to lower management, and then to employees in general. In an organization with low morale, sticking

to the rules to protect your career may be better than breaking them even if the rules are inappropriate. This defensive posture might spread to allied agencies, such as the Transportation Security Administration, which is already less concerned with safe transit than terrorists' potential to use transportation as a weapon. A hypothetical situation could prompt this question: Is the TSA official in charge of the security of a local airport very likely to tell his employees to stop doing their principal job and just let the evacuees through? Not if he knows that FEMA officials are not sending water and food to the airport because airport staff cannot send the proper requisitions because the fax machines are out of order. The message may be that in perilous times it is best to go by the book. (While not unreasonable, this is not substantiated by research that I am aware of). This is a different explanation than "they panicked," or "the storm was so large and the task so unprecedented."

A further consideration is that the reorganization of FEMA into the Department of Homeland Security imposed a top-down, command-and-control model on an agency that most experts say should maximize the power of those at the bottom. The centralization would reinforce a tendency to go by the rules even if the situation suggests they are inappropriate. Maximizing the ability of the lowest level to extemporize and innovate will minimize the bureaucratic responses that so characterized FEMA. A frequent criticism of the reorganized FEMA was that the centralized DHS model, and the removal of authority for preparedness to other parts of DHS, would inhibit its responsiveness to unique events. (Glenn 2005) (Even worse is a centralized organization with leaders that ignore the call to arms.)

The failures of the National Guard are harder to account for. The head of the National Guard Engineer Battalion hiding in the convention center ordered, and got, more ammunition and barricaded the entrances. He admitted they could have gotten the center under control, if so ordered, but senior commanders ruled out that possibility. (Haygood and Tyson 2005) So they left in the night. Were senior commanders taking cues from the top and deciding not to bend the rules? Could it be that in the absence of an

energized response from the White House, the heads of state guard units and of the military in Washington did not feel they should tell their units that this was a disaster of such magnitude that they should use their best local judgment? The heads of military units near New Orleans awaited orders to help that never came, as did a naval hospital ship in the gulf. How can we explain the many cases elsewhere during the crisis where the guard and the regular military performed well? Were they cut off from their commanders, and this allowed them to innovate and respond?

This account of the failings of the DHS in the instance of the Katrina Hurricane relies heavily on the characteristics of the Bush administration. But if the United States is to protect itself from devastating hurricanes, we must envision the possibilities of having more administrations such as this one. Just as organizations must be structured such that even with leadership that is just average (and even below average, as one cannot rely on having above-average leadership any more than the fictional town of Lake Wobegon can expect only above-average children), so must we prepare for organizational failures. The best preparation for hurricanes is to prevent the concentration of populations in risky areas. Even under the best of administrations, the devastations from a storm such as Katrina will be enormous.

## THE DHS AND INTELLIGENCE

Most commentators see the biggest failing of the largest reorganization of the government in recent times—the creation of the DHS—as the failure to connect it with the intelligence agencies. The hub of DHS's dot-connecting efforts was to be a new intelligence center for tracking terrorists. Just four days after Ridge was sworn in as the first secretary of homeland security, he found that DHS would not control this center. "Ridge and his aides thought the center was one of the key reasons the department had been created, to prevent the coordination failures that helped produce Sept. 11. Not only had the White House undercut Ridge, it also let him

find out about his defeat on television." (Glasser and Grunwald 2005) The importance of this decision cannot be overemphasized.

The new agency was no match for the agencies in the intelligence community (IC). The Department of Homeland Security only managed to get one office of the FBI (the National Infrastructure Protection Center) and assurances that a few DHS members could sit in on the coordinating committees in the IC. It can ask for information but has no assurance it will get it. Since intelligence is critical for security, for deciding where to put resources, for information on what kind of threat is likely, for alerts that a threat is imminent, for knowledge of the strengths and weaknesses of terrorist groups, and so on, the DHS is almost totally dependent on an intelligence system that is not decentralized but fragmented. Of course, many things are obvious: cockpit doors of aircraft must be hardened; container ports are vulnerable, as are national landmarks and nuclear power plants and chemical plants, among others. But intelligence is needed to decide how much money and effort should go to each type of target, since perhaps only a third of the funds needed to do the most obvious things are available. What are terrorists most likely to attack?

Even when the DHS gets warnings from the IC it has had trouble communicating with its own agencies. Both federal and state agencies said they were informed of "orange alerts" only by watching CNN, not through notification by the DHS, which issued them. And some governors and mayors refused to respond to the orange alerts since they were so vague and response was so expensive. The DHS appears to find it difficult to be responsive itself. Testing the capabilities of its police force, the U.S. Park Police deliberately left a suspicious black bag on the grounds of the Washington Monument. The police failed to respond quickly or effectively. One officer reportedly was caught sleeping. When a test official called the Department of Homeland Security to warn them about the bag, he got this priceless recording from our protectors: "Due to the high level of interest in the new department, all of our lines are busy. However, your call is important to us and we encourage you to call back soon." (Shernkman 2004)

Should the DHS have gotten more control over those security agencies that are not clearly related to military strategy and battle-field tactics? (The bulk of the estimated $40 billion spent yearly on security is military-related.) Should it have at least gotten the FBI, which is primarily concerned with domestic security, though it does operate abroad to some extent? Aside from the problem of increasing the sheer size of the DHS even more, the consensus is that the security agencies were far too powerful for even parts of them to be moved. As Amy Zegart argues, agencies concerned with foreign affairs, such as the intelligence agencies, are oriented toward the president, rather than Congress, and controlled by the president to a greater degree than domestic agencies. (Zegart 1999) The interests of the president and of the intelligence agencies thus were "aligned," as political scientists put it. Even if it were wise to give the nonmilitary intelligence agencies to the DHS, that was not likely to happen if the DHS could not even get control of the FBI. Worse yet, while DHS had the statutory responsibility for establishing a common watch list out of the twelve that existed, even this task was given to the Justice Department, over the objections of Clark Kent Ervin, its beleaguered inspector general. Ervin has a graphic chapter on the DHS's failure to do the main job it was set up to do: coordinating disparate intelligence efforts. (Ervin 2006; Mintz 2004)

Since the DHS was unable to coordinate the efforts of the twenty-two diverse agencies over which it was given nominal control, it was highly unlikely to be able to coordinate the efforts of those agencies over which it had no control. Regardless of the structural reasons that would seem to make it logical for the agency to incorporate the FBI, it is not a happy thought. Quality leadership at the top of the DHS was not available, and the insular FBI would resist the kind of incorporation that would guarantee its cooperation with DHS headquarters: most of its focus remained on drug interdiction and nonterrorist criminal activity; it would have little or nothing to do with natural and industrial disasters; and it had powerful friends and constituencies in Congress and law enforcement agencies that would resist changes. But co-

ordination of intelligence outside of the DHS appears to be just as difficult.

When the 9/11 Commission released its report in the summer of 2004 (Staff 2004), it triggered another burst of government reorganization. The commission recommended a radical change in the intelligence community. It was to be headed by an intelligence director with cabinet-level status and the authority to determine the budgets and key personnel of all of the fifteen agencies that made up the IC. The commission was on well-trodden ground with its recommendation. In just the past ten years there have been thirteen major studies and reports concerning our national intelligence system. They all recommended reorganizations, particularly to centralize controls over the disparate activities. Political scientist Thomas H. Hammond asks, "Why is the intelligence community so difficult to redesign?" (Hammond 2004) Bureaucratic politics and power of the sort we have been examining play a role, he admits, but there are more basic structural reasons.

Hammond argues that because of dilemmas that are inherent in any structural set up, any reorganization plans are bound to have enough faults in them to prevent any agreement on basic changes. For example, the intelligence community both collects information and integrates and disseminates it. A structure that is good for collection may be poor for integration and dissemination, and vice versa. Furthermore, a structure that favors rapidly acting on intelligence in any situation short of an imminent attack—say, acting on the August 2001 warnings and information about flight schools, etcetera—has costs. It may disrupt the source of information and prevent further surveillance that could identify more terrorists and their organizations. (This is the classic tension between the FBI and the CIA.) Finally, a structure that is appropriate for dealing with one kind of threat, for example, state-sponsored terrorism, will not be appropriate for another kind of threat, such as that presented by Al Qaeda. A major criticism of the Bush administration's handling of terrorism from 2000 to 2001 was that it was still preoccupied with state-sponsored threats from North Korea, Iran, Syria, and even to some extent Russia, whereas the IC should

have been reorganized to deal with the mounting threat of Islamic terrorist organizations.

But neither structure might be appropriate for domestic terrorism as represented by the Oklahoma City bombing, by leaderless groups, or by individuals loosely connected for an action on abortion clinics, power grids, logging operation, or animal rights—we have had terrorists in all of these—and then dissolving. All three forms of terrorism, state-sponsored, foreign jihadists, and domestics, are still present dangers, but we can hardly have three separate structures to deal with them. The appearance of non-state-sponsored groups like Al Qaeda has not meant the disappearance of state-sponsored terrorism or even of anthrax mailings. So how should the structure be oriented?

The complex trade-offs required have produced a kind of structural conservatism on the part of intelligence policy makers, Hammond argues. No alternative structure has seemed clearly superior to the present one. And of course the costs of tearing organizations apart and disrupting career paths are substantial.

Hammond illustrates the difficulties when he outlines six functions and policy areas of the intelligence community, and they give pause to any easy solution. Recognition of the variety (and possible incompatibility) of these functions and policy areas does not seem to have been addressed by the December 2004 intelligence reform bill signed by the president. They are:

- Determining the intentions and monitoring the capabilities of the former Soviet Union, China, North Korea, Iran, etcetera
- Monitoring the nuclear proliferation technologies, capabilities, and intentions of foreign state and nonstate organizations
- Counteracting terrorism at home and abroad
- Providing intelligence support for antidrug campaigns
- Providing intelligence support for U.S. government policy in Iraq
- Supporting U.S. combat operations

The intelligence reorganization of December 2004 primarily involved the creation of a new director of intelligence that has budgetary control over all fifteen agencies. This prompted strong opposition from the Pentagon, which was eventually defeated when the public outcry finally brought the reluctant president to force a congressional committee chairman allied to the Pentagon to back down. (Another chairman made the issue of illegal immigrants hostage to the bill but only partially succeeded.) The details of the reorganization and budgetary changes were "still to be determined" at the end of 2004, and as of this writing, Spring 2006, it is far from clear what will emerge from the rather vague legislation. It does not appear to address the dilemmas that Hammond identifies.

Senator Robert Byrd (D-WV), one of only two members of the Senate to vote against the bill, delivered a scathing speech dealing with the limitations of the bill. Congress in general wished to give more powers to the head of the new agency, but the Pentagon and other interests got the powers watered down. However, Byrd was primarily concerned about the hasty passage of a long, complicated bill whose latest version the Senate had only twenty-four hours to review: the secrecy the bill provided to the new agency, closing off ombudsman reviews and whistle-blower protection, the failure to deal with prison scandals associated with intelligence interrogation and the successful attempt to limit inquiries into possible prison abuses; the change from mandates to promises regarding new resources; and the successful attempt to reinstate and reinforce powers under the Patriot Act that Congress wanted reconsidered, parts of which the courts had thrown out. The intelligence bill was described by civil liberties groups as a Trojan horse, using the opportunity for reform of intelligence failures to greatly weaken civil liberties. (See, for example, Eggan 2004.) Senator Byrd's remarks touch on many of the organizational issues we have been dealing with. It is a sobering litany of why we should not expect much from the government organizations that are supposed to protect us and the uses to which our organizational efforts might actually be put. (Byrd 2004)

An impressive lineup of experts argued in vain against hasty consideration. Byrd pointed to the following flaws, among others: the thoroughly politically partisan Office of Management and Budget can screen intelligence testimony before it is presented to Congress; the requirement for inspector general and ombudsman positions was removed, and it was left to the discretion of the intelligence director whether to appoint them and what their powers would be; an independent Civil Liberties Board, recommended by the 9/11 Commission, was made dependent on the Office of the President; the Foreign Intelligence Surveillance Act was being modified to give the executive branch power to undertake electronic surveillance, allowing the president to monitor domestic telephone calls.

The problem of congressional oversight remained after the reorganization of intelligence. An attempt to establish a single supercommittee in each branch of Congress fell prey to the interests of existing committee chairmen. The chairman of the Rules Committee of the House complained that giving jurisdiction over the Transportation Security Administration (newly established) and the border control to a supercommittee left him with "scars." "I will be dining alone," he said. Another representative, Curt Weldon (R-PA), said, "but when you read the legislative language, it guts all the authority," leaving it in the president's hands, which was one of Senator Byrd's points. Another Republican representative said: "I think we're fighting tonight for the soul of Congress. It's turf battles, it's people who want to go back to September 10" in terms of congressional oversight. (Kady 2005)

I think that the most important factor will be the intentions of the president. He has appointed a head that presumably shares his political vision, but the Pentagon is suspected of starting its own domestic intelligence agency that will be out of reach of the new director. Structural changes are needed, and one may rejoice that a terrorist threat integration center will be a substantial part of the new agency. But the new head of the IC is unlikely to solve the dilemmas, even with a new terrorist threat center. A February 2006 news story reported that lawmakers were worried that the director of national intelligence, John Negroponte, had not moved quickly

enough to establish his leadership and had not been able to exert effective control over the Pentagon. It was also feared that rather than being a lean operation, the agency was becoming another bureaucratic layer. (Pincus 2006) A terrorist threat integration center established in the Clinton administration was moderately successful; the same center in the Bush administration was not. A very great deal depends on executive leadership, much more than structural reorganizations, as important as they may be. It is an observation I have made in connection with the diversion of the goals of FEMA in the Reagan administration, and will see again with the Millstone nuclear plant failures and others in future chapters.

## DREARY CONCLUSIONS

There is no doubt in my mind that the nation is somewhat safer since the 9/11 attack. Suspects have been apprehended, the Federal Aviation Administration has made changes, so has Immigration and Customs Enforcement. But the first two improvements were made outside of the new department of Homeland Security, and the third easily could have been without its appearance. As we shall see in a later chapter, the department has had very limited success in making our chemical and nuclear piles of vulnerability more secure. Our borders are still so porous that it would be sheer luck if a guard happened onto a terrorist. A few of the thousands of containers that daily enter our ports are said to be under some surveillance, and the department has been active there but has been thwarted by large shippers, principally Wal-Mart, which has given $191,500 to current House Homeland Security Committee members since 2000 (one of the advantages of being on such a committee). One of Wal-Mart's friends, the National Customs Brokers and Freight Forwarders Association, put it bluntly: "The private sector needs to continue to get emerging government figures to swear on a stack of Bibles that commercial operations are an important responsibility that cannot be subordinated wholly to security interests." (AFL-CIO 2006) Barry Lynn provides a detailed ac-

count of Wal-Mart's opposition to strong port-security legislation. (Lynn 2006) But the new surveillance (and more breaches of basic privacy, unfortunately) of populations that might harbor terrorists is handled by Justice. Billions have been spent to improve intelligence and first responder capabilities, but intelligence funding is outside of the Department of Homeland Security. That does not leave us with much to be grateful from the department.

And we have no idea how many more billions would need to be spent, and where to spend them, in order to close all the holes in our open society. I think it is foolish to think our society will ever be safe from determined terrorists, but it is possible that we have raised the bar just enough to make it a bit more difficult for them, and this may be at least a small part of the explanation as to why we have not been successfully attacked on our soil since September 11, 2001—more than five years at this time of writing. (Better explanations for the hiatus on attacks will be explored in the last chapter. In essence, they argue that the United States has been shown to be vulnerable, and that is enough. There is room for small attacks, of course, but more pressing concerns for the Islamic extremists are getting "infidel" troops out of Islamic nations and destabilizing Islamic regimes that are "corrupted" and shaky by driving out all infidels and installing fundamentalist regimes.)

Our efforts have, in some cases, made us safer from our two other sources of disaster, natural and industrial ones. Certainly, strengthening first-responder capabilities will often mitigate natural and industrial disasters. We need better intelligence for these types of incidents as well as for terrorist attacks, improved medical response to epidemics as well as biological attacks, and training and simulated emergencies for all three threats. But there is a danger in the "all hazards" justification for the structure of the DHS. The department focuses on terrorism, and the most expensive parts of its program will have little to do with industrial or natural disasters: upgrades for police departments are favored over those for fire departments; coping with biochemical attacks receives substantial funding, but not coping with epidemics, which are much more likely to occur and more devastating; intelligence and sur-

veillance activities (watch lists, border security, cameras in subways and on streets, surveillance of antiwar groups and mosques, etc.) are extensively funded but have nothing to do with natural disasters. No one has done an accounting that I am aware of, but I suspect that most of the money tries to protects us from only one of our three disaster sources.

But we have a porous society, less protected (and less inconvenienced) than those of our European allies and Israel. A few suicide bombers coordinated to blow up tunnels, bridges, and airports would panic our government, and domestic or foreign terrorists are capable of shutting down the Northeast power grid for weeks with a few well-placed, small explosions. Suitcase bombs in a chemical plant (which are still unprotected, as we shall see in Chapter 6) could put seven million people at risk if the weather cooperated. A drive-by attack on the spent-fuel cooling tank at a nuclear power station would be harder to pull off but quite possible and could release more radiation than is held in the core in minutes. These are targets of our own making, made large and vulnerable for reasons of small economies and unwillingness to have a few inconveniences. Very little is being done, and little can be done, about all three threats, other than basic reductions of our vulnerabilities. We are unprepared, and should not be prepared, to have perhaps 15 percent of our employed population engaged in protecting our infrastructure. With our form of federalized government and the problems of campaign financing and corporate power, government can only muddle through its large and its small problems, as Charles Lindblom argued almost fifty years ago, including its new problem of a Department of Homeland Security. (Lindblom 1959) Still, it is possible that the criteria by which I and others have judged the first five years of the department may set too high a bar. Perhaps the bar should be "Have we made terrorists, even suicidal ones, pause just a bit, maybe for five years, before trying more attacks?"

Nevertheless, it is useful to analyze the supposed failures of the DHS, as well as its troubled origins and those of the intelligence community. We can always do a bit better. And through this analy-

sis, we can better understand what we are up against when we organize and reorganize. It makes the plea to reduce our basic vulnerabilities all the more compelling. Such a reduction would also be a very difficult project, perhaps more difficult than (and almost as improbable as) a reasonable defense against terrorism, but I think not. As we proceed through other disaster-prone aspects of our society in this book, I will slowly build the case for such a reduction.

# PART 3

## The Disastrous Private Sector

----------------------------------

# 5 Are Terrorists as Dangerous as Management?

## The Nuclear Plant Threat

------------------------------------

NUCLEAR PLANTS IN THE United States present two sources of cataclysmic danger. One is stored nuclear waste products, planned for Yucca Flats in Nevada, which threaten to contaminate vital water supplies. Given the fearsome predictions associated with global warming, the area may be unsuitable for agriculture in one hundred years anyway. More fearsome in immediate terms is the release of radiation from one of our 103 operating plants because of natural disasters, industrial accidents, or terrorist attacks. Tens of thousands of people might die and land equivalent to half of Pennsylvania become uninhabitable. Terrorists could do this right now with simple weapons. An industrial accident could do it; we have come very close to meltdowns several times in the last three decades. Most likely, the top management in the utility companies and in the plants could bring it about through neglect of maintenance and safety rules, sometimes willful neglect. Two themes will by now be familiar: the failure of regulation in an age of privatization and the downsizing of government; and the inevitable, prosaic failure of organizations. A new one makes its appearance here: industrial concentration. We will examine it in this and the next two

chapters. In the case of nuclear power plants, I argue that the willful neglect of safety suggested in two case studies and documented in a third is the result of the consolidation of the electric power industry, magnifying the vulnerability of the bottom line. We cannot expect to downsize or deconcentrate the nuclear power plant itself, but we can do a great deal to make it much safer through better regulation and industry deconcentration. Much safer nuclear power plants can perhaps be built, but I am concerned with the ones operating now, most of which are receiving authorization to extend their plant life by twenty to thirty years.

## TERRORISM AND NUCLEAR PLANT SAFETY

In 2000, in St. Petersburg, Florida, the leader of a cell of the Southeastern States Alliance, a radical antigovernment group, was arrested and charged with possessing a cache of arms and planning to take out the nearby Crystal River nuclear plant with a "strike team" of thirty confederates. Presumably he was a domestic terrorist; his motives were unclear but his plan was simple and could have worked. The team planned to use explosives to disable the power grid that fed power to the plant. It is not comforting to learn that the explosives were found to have been stolen from the National Guard armory, where the leader was a regimental commander. (Leisner 2000)

Emergency power at nuclear plants is provided by diesel generators (which have a long history of failing to start and other problems). Clearly visible in some plants, these generators could be taken out with grenades. Or, a hurricane could do the work of the terrorists' dynamite and take out the power, and the storm could easily render the emergency generators inoperative as well. Or there could be a simple power failure. New York State's Nine Mile Point Nuclear Station had a sudden power failure in 1996, knocking out vital instruments and warning lights. The backup power generators also failed, and operators were unable to monitor the reactor core for twenty chilling minutes. It was, a state commis-

sioner said, like going seventy miles an hour down a road at night
and losing your speedometer, dashboard lights, and headlights. He
did not mention that vastly more people were at risk than those in
the car. (Staff 1991a)

Disrupting the power supply of the plant is not even necessary if
one knows about safety equipment at the plants, which could be
gleaned from the diagrams found in Al Qaeda caves in Afghanistan. According to U.S. intelligence, there was "pretty convincing
evidence" that Al Qaeda operatives had been "casing" U.S. nuclear
plants before the September 11 attacks. (Borenstein 2002; Sagan
2003) U.S. intelligence agencies issued a warning in January 2002
of a potential attack on U.S. nuclear plants and government nuclear
facilities. (Gertz 2002) As a *U.S. News & World Report* story notes,
the water supply system needed to cool the reactor core, the emergency generators, and vital controls essential to the plant's safety
are all within easy reach of an attacker who does not even have to
enter the plant's perimeter. Our government's Sandia Laboratories
speculates that a truck bomb outside the (thin) gate could produce
catastrophic radiation releases. (Pasternak 2001) Nor are the
doors locked. Scott Sagan, in his detailed account of security failures, quotes news accounts to the effect, for example, that it took
the Department of Energy (DOE) thirty-five months to write a
work order to replace broken locks at a weapons lab facility and
forty-five months to correct a broken doorknob that was sticking
open and allowing access to sensitive sites. (Sagan 2003) Nor is the
White House Office of Management and Budget very cooperative:
the DOE requested $138 million in emergency funds to improve
the security of weapons and radioactive wastes soon after 9/11, but
the OMB rejected 93 percent of the funds. (Sagan 2003)

The spent-fuel storage pools are a particularly attractive target.
At one-third of our plants, the pools of frigid water holding the
highly radioactive spent fuel rods are outside the main building
and vulnerable. If they are not continually cooled, the water will
boil off in a matter of two to five days and the radioactive rods will
go off like sparklers on the Fourth of July, potentially spreading
more radiation than a core meltdown might. The spent-fuel stor-

age pools need outside power to keep cool if the plant fails; the most immediate sources are the emergency diesel generators, but those will run out of fuel in a few hours. If the reactor is leaking radioactive particles, no one could get close enough to refuel the generators. An accident, a terrorist attack, or a severe weather impact could prevent power coming in from the electric power grid. Even without a shutdown of the reactor, a suitcase bomb could disable controls or rupture water supply pipes; a large suitcase bomb could blow a hole in the side of the huge pool, causing the cooling water to disappear, and radioactive releases would begin almost immediately. The highly respected Robert Alverez writes in the *Bulletin of Atomic Scientists* that this is the single most fearsome vulnerability of nuclear plants. (Alverez 2002) A panel of the National Academy of Sciences is equally worried. (Wald 2005b)

In 2001, the nuclear industry was not worried about the vulnerability of nuclear plants to terrorist attacks. Indeed, a spokesperson for the industry's trade group, the Nuclear Industry Institute, said, "We believe the plants are overly defended at a level that is not at all commensurate with the risk." (Alverez 2002, 44) However, overly defended or not, there have been attempted attacks in the United States and other nations. In 2000, Japanese police arrested a man with seven pipe bombs who was planning to blow up a uranium-processing plant; in September 2000, a group planning to sabotage the functioning reactor at Chernobyl was apprehended. In the United States, there were at least thirty threats against nuclear plants between 1978 and 2000. In 1989, four members of the radical environmental movement, Earth First, were charged with conspiring to disable three of the four power lines leading to the Palo Verde nuclear power stations in Arizona. (Wald 2001)

Of course, since all these attempts failed, one might say the plants are "overly defended," as the industry did. Apparently the Nuclear Regulatory Commission (NRC) agreed, because in the fall of 2001, it introduced a new program that *reduced* federal oversight of security and allowed the power companies to design their own security exercises, despite reviews that found, in 2000,

"alarms and video camera surveillance cameras that don't work, guards who can't operate their weapons, and guns that don't shoot." (Pasternak 2001) Plant-security plans and their exercises are not encouraging. Several organizations have done studies on the security issue, including the Government Accountability Office and various environmental organizations. Here are some of the disturbing findings:

- The first warning of vulnerability came a few days before terrorists first attacked the World Trade Center in 1993. A former mental patient drove past a guard shack at the "overly defended" Three Mile Island facility, crashed his station wagon through a metal door and managed to drive sixty feet *inside the turbine hall*. He then fled on foot and it took a *few hours* to find him. He was unarmed. Security experts observed that a truck with a bomb the size of the one that so severely damaged the World Trade Center could have been detonated at the gate of a nuclear power plant and caused a major radiation calamity. The NRC promised tighter security. (Wald 2002) But as we have seen, it reversed itself in 2001.
- Half of America's ten nuclear weapons research and production facilities had failed recent security drills as of October 2001. The U.S. Army and Navy commando teams were able to cart away enough weapons-grade uranium to build several nuclear weapons in three cases. (Hedges and Zeleny 2001)
- The GAO reported in 2003 that security guards failed to search people who triggered alarms when going through metal detectors. During exercises to test security, plants used more personnel to defend the plant than are available on a normal day. Unrealistic rubber guns were utilized in these required exercises. (GAO 2003)
- Despite steady reassurance by the NRC that the security system has been upgraded since 9/11, a survey of twenty guards in thirteen plants by an government watchdog

group, Project on Government Oversight (POGO), found that in three-quarters of the plants examined, personnel were not confident that an attack could be defeated. The guards said that morale was low; they are paid less than custodians or janitors; their training is far less than regulation requires; and they do not have the automatic weapons and sniper rifles the terrorists would have. If a terrorist with a backpack of explosives jumped the fence and headed for the pump house, the spent-fuel storage pool, the backup generators, or the reactor itself, the guard could only observe and report the event. Exercises run by the NRC found it would take one to two hours for outside responders to arrive with a SWAT team, while a successful attack would be over in three to ten minutes. (POGO 2002)

- Mock attacks, for training and certification purposes, until 2002, employed three attackers, whereas actual terrorists would use an estimated ten to twelve. (Four or five attackers were used in a drill at Indian Point after 9/11, which a member of the NRC declared a great success but POGO ridiculed, and Senator Charles E. Schumer [D-NY] agreed with POGO. Wald 2003a) Speaking of government nuclear installations (and speaking in its usual tortured prose), the GAO announced in 2004, "While the May 2003 DBT [design-basis threat] identifies a larger terrorist threat than did the previous DBT, the threat identified in the new DBT, in most cases, is less than the threat identified in the intelligence community's Postulated Threat, on which the DBT has been traditionally based." (GAO 2004d) That is, the experts expect much larger attacks than the plants train for. In some cases, warned of the exercise, the plants had more people available to repel invaders than would work there in a normal shift. Not until February 2002 did the NRC require guards to carry their "primary" weapons (i.e., shotgun or rifle) when on duty. If anything looked amiss, they had been expected to trot over to a building where their equipment was stored in lockers. The utilities can choose

the date of the mock attack, the kind of attack, and the attackers (local police or even utility management or training staff); they can script the place of entry and plan of attack; and they can have their security personnel carry the communications equipment and bullet-proof vests, which normally are stored away, on the day of the attack. Attacks are in daylight, and the three "terrorists" doing the attacking have limited weapons. This "dumbing down" of military exercises seems even too tough for the utilities, since nearly half fail. For example, a mock terrorist took a badge from a guard declared dead by those supervising the test and used the badge to enter a building unchallenged. The utility then complained to the NRC commissioners that this was cheating because such a tactic had not been scripted. (POGO 2002)

- Privatizing security drills at regulated sites with catastrophic potential is risky. The contractors conducting the drills of the security personnel they supply to the plant have no incentive to make them tough; they might lose their contract. The plant does not want either the embarrassment of failing to repel the attackers or the added expense of improving its security. And the NRC does not wish to be charged with inadequate regulations and oversight. No one wants failure, so the incentives for the exercises to be unrealistic are great. This became clear in the case of the Oak Ridge nuclear plant that makes warheads, where an investigation disclosed cheating in the mock attacks over a period of two decades. Barricades were set up to alter the outcome; guards deviated from their response plan to improve their performance; guard supervisors from the private company, Wackenhut Corporation, were allowed to see computer simulations the day before the scheduled attacks; guards were improperly told which buildings would be attacked, the exact number of attackers, and that a diversion was being staged. Attackers used guns that send a laser beam, and guards have sensors to see if they have been hit.

But the guards disabled their sensors and disabled the weapons used by the attackers beforehand, resulting in a high score for Wackenhut Corporation, the largest supplier of guards for U.S. nuclear facilities. (Mansfield 2004)

The terrorism picture, the third of our three major sources of vulnerability, is much as we would expect (and much as we shall find in the case of another major source of our society's basic vulnerability: chemical plants). The nuclear power industry denies there is a serious problem ("plants are overly defended now"); the major regulatory agency, the NRC, is hardly mobilized; there are no unscheduled tests; and the implementation of safety programs is comically inept, unmotivated, and even corrupted by safety vendors such as Wackenhut. Government installations concerned with nuclear weapons, which we have only briefly mentioned, appear to be no better in these respects. For society's most concentrated source of destruction, this is a gloomy picture. As we shall now see, another one of our three sources of vulnerability—accidents in industrial organizations—is not much different.

## NUCLEAR PLANTS AND OPERATIONAL SAFETY

When nuclear energy first appeared, the untried and untested plants had a rocky time. There were meltdowns or explosions or serious fires in four U.S. plants within a short time of their start-up. An experimental sodium reactor, the SRE, had a partial meltdown after fourteen months of operation (1959). Commercial plants failed even sooner. The Fermi plant had a partial meltdown within two months of its opening (1966), despite scientists' claims that it would be impossible; the Three Mile Island plant had a partial meltdown of one-half of the core within three months (1979); the St. Laurent des Eaux A1 plant in France had a partial meltdown within four months (1969); the Browns Ferry plant in Alabama had a serious fire after six months (1974); and the Chernobyl reactor managed to go for one year and seven months before its 1986

melt down that killed as many as 32,000 people by some estimates. (Lochbaum 2004, 5–6) (Some people disputed this estimate, but more important is the luck that the USSR had with the weather that night. The radioactive cloud rose slowly to several thousand feet in the still air before dispersing. Had a wind carried it over nearby Kiev, many more thousands could have died.) Things were so bad in the industry that some plants were never completed, or were completed but never brought online. The NRC itself acknowledged its contribution to the problem in a 1984 report that said there was an inability to ensure adequate control over design and construction and an inability to implement quality assurance controls; that the NRC itself "made a tacit but incorrect assumption that there was any uniform level of industry and licensee competence"; and that the limited inspection resources of the NRC meant that there was inadequate inspection of the design process. (Lochbaum 2004, 10) In short, the emphasis was on getting the plants running, regardless of the construction failures and the basic design failures.

In the United States, the nuclear-plant building program came to an abrupt halt when the Three Mile Island accident revealed the dangers of this philosophy. Fully twenty-seven of the one hundred or so nuclear power reactors have been shut down since 1984 for more than a year for extensive repairs to safety equipment. The year-plus durations of these shutdowns are prima facie evidence that problem identification and resolution programs at these facilities were seriously flawed if not quite totally dysfunctional. Years of overlooking problems and applying "Band-aid" fixes at these plants resulted in a backlog of safety problems that took a long time to resolve. Effective problem identification and resolution programs could save plant operators time and save money in the long run, argues a Union of Concerned Scientists report. (Lochbaum 2004)

With plants designed for twenty years of operation asking to be allowed to extend their life by another twenty years, the problem of aging parts and vessels is mounting. The Union of Concerned Scientists, in reports written by David Lochbaum, considers aging

facilities to be potentially a very serious problem. Our database of accidents cannot tell us how serious the aging might be. We don't even have very good evidence as to whether the reliability of our plants is stable, going up, or going down. Numerous measures are available; but they point in different directions and are hard to interpret. Perhaps the most clear and simple one is the number of serious incidents (i.e., near misses). Here the news is mixed. The number dropped from 0.32 per reactor year in 1988 to a low of 0.04 in 1997, perhaps reflecting a maturing of the industry and few new plants coming online. But it then rose to 0.213 in 2001—about twenty-one serious incidents a year if there are one hundred plants operating all year—perhaps reflecting the aging problem. (Lochbaum 2004, 19)

The problem with this and other statistical measures is that they emphasize the *probabilities* of failures, whereas with systems that have catastrophic potential, such as nuclear plants, we are also interested in *possibilities*. (Clarke 2005) The industry and many academic nuclear scientists point out that no one has been killed by a plant failure in the United States, so they say the probability of future deaths is vanishingly low. Environmental groups and many academic social scientists argue that we have given the small number of plants running only a brief chance to reveal their fury—only two or three decades. More important, the latter argue, it is what is possible that counts with such a deadly technology. As the old bumper sticker sardonically put it, "a nuclear plant meltdown can ruin your whole day."

Nuclear plants occupy a special place in our industrial landscape because their catastrophic potential is so huge that they must be heavily regulated, but their complexity makes regulation difficult (there are so many rules required, and so little experience with all the possible interactions and conditions). It also makes regulation burdensome and intrusive: endless forms and reports are required, and inspectors are constantly on site. European plants have a better safety record because the plants are either state-owned, with safety built in at the start more effectively than the privately built and run U.S. plants, or if private they are chartered by the state,

with a less adversarial relation between the government and the private owners in Europe. Our adversarial culture, reflected in our labor relations and regulatory activities, can be attributed to our free-market capitalism. It fosters a short-run perspective and cost-cutting to increase profits. (See the excellent discussion of the difference between U.S. and European plants and regulations in Jasper 1990.) A third model is the former Soviet Union, where the state owned everything and the emphasis was on meeting unrealistic targets by neglecting sound engineering principles and safety. (The fullest account of the Chernobyl accident and the USSR nuclear power program—an alarming account—is by Sonja Schmid [2005].) With deregulation of the energy industry in general, our plants are subject to more intense competition with each other (electricity is now shipped much further over the transmission system), more competition from more traditional sources of energy (which can produce power more cheaply; when all costs are considered nuclear power is much more expensive than other source of power), and widespread consolidation of generating plants and accompanying downsizing of employees and longer workweeks. But even where the state regulation is strong, as in Japan, there will always be an incentive to cheat.

Japan's nuclear plants for some decades appeared to perform at higher levels of efficiency and safety than those in the United States. But in July 2000, four ominous unexpected shutdowns occurred, some releasing unacceptable radiation levels, in the plants run by Tokyo Electric Power Company (TEPCO), Japan's largest utility. In 2001, a whistle-blower triggered disclosures of falsified tests at some of the company's seventeen plants, and the government forced TEPCO to close some plants. In 2002, the company predicted that *all* of its seventeen plants might have to be shut down for inspection and repairs, because of falsified inspections and concealment of faults found in inspections that the government ordered; some of the faults were potentially catastrophic. A top company official was charged with giving specific orders to hide large cracks in the "shrouds," or steel casings around the reactor core, in two of the thirteen reactors at which false inspection reports had

been filed. (Later I will designate this as an "executive failure," as distinguished from an organizational failure, when I discuss similar orders in the Millstone nuclear power facility.) The president and vice president of the company resigned, the salaries of three executives were reduced by 30 percent, and those of fifteen others were cut. (Radsafe 2004) I know of no resignations or salary cuts following similar disclosure of executive failure in the United States. (State ownership by itself does not guarantee safe practices; it has not in either Russia or Japan. It seems to have worked quite well in Europe. Private ownership with poor state regulation seems less effective than European state ownership.)

The U.S. economic system favors short-run indicators and mobile capital flows. Managers are both creating this culture and being driven by it. Thus, the short-run savings that accumulate with cutting corners on maintenance and safety can be expected to dominate management thinking at the top, middle, and bottom. Since any untoward consequences of short-run savings are unlikely to appear, if they ever do, until the distant future, management can escape accountability. This has to be expected as a risk that most large organizations will take, and those with the most market power will be able to more easily absorb the consequences or deflect criticism. The organizations supplying our deregulated power are now larger and have more market power, as we shall see in chapter 7 on our power grid.

## THE INDIAN POINT, DAVIS-BESSE, AND MILLSTONE EXAMPLES

Our first example, the Indian Point power plant, illustrates the problems confronting what was rated as one of the five worst plants in the United States. We don't know enough about the role of top management in all these problems, so they may have more to do with failures by operators or managers, lack of skilled personnel or lack of money, or perhaps even technological uncertainties, though that seems unlikely. It will give us the flavor of re-

peated problems that even the better ones face. It is not encouraging.

The second example, Davis-Besse, presents a case of repeated warnings by engineers that were countermanded by top management, bringing the plant very close to a serious meltdown, along with all the dispersion of deadly radiation that would accompany this. We will also get a closer glance at the ineffectual Nuclear Regulatory Commission, which worried mightily and frequently about safety at this plant, but could not intervene effectively.

The Millstone plant suffered the range of failures of the other two plants but, even more than the Davis-Besse example, it locates conclusively the source of this failure: top management. Organizational theory has not done a good job in distinguishing between the failures of workers or management—part of the difficulty of getting organizations to perform well—and deliberate, knowing malfeasance on the part of executives. This is largely because we rarely have the data to conclusively make the distinction. We are able to document this example of what I will call *executive failure* because of the extraordinary work of a state public agency. The Millstone plant also came close to a major disaster.

### Indian Point

For the first of three examples of our risky nuclear plants, I will draw upon Elizabeth Kolbert's account in the *New Yorker* of the Indian Point plant. (Kolbert 2003) The Indian Point nuclear power station is not the worst in the nation, but it is probably typical of the operational problems that these plants still face, including problems due to aging. It is unusual, however, because it is so near to New York City. Half of all nuclear plants are near urban areas, but this is a particularly large urban area with no adequate means of escape. More than twenty million people live within fifty miles of the plant and, depending upon the wind, many millions of these could be severely harmed, and those close at hand, killed. The 300,000 most at risk are in the ten-mile "emergency planning zone," but

there could be thousands of deaths among the several thousand who live within seventeen and a half miles, the "peak facility zone." The devastation that the old Atomic Energy Committee said could result from a meltdown at Pennsylvania's Three Mile Island plant— contaminating an area half the size of Pennsylvania—applies to the New York metropolitan area and much beyond.

We pick up the story in 1992 when one of the two reactors was placed on the watch list because of long list of safety lapses in previous years. In the most recent lapse, in 1991, the backup safety system for shutting down the reactor in case the primary one failed, a system essential for safety, was found to have been inoperative for six months. The plant was given a small fine: $225,000. But shortly after the fine, a problem was found with a set of valves, and the engineers rushed to replace them before an upcoming NRC inspection occurred. They put them in backwards, blocking the essential cooling system, and had to shut down the plant.

Under NRC pressure, the New York Power Authority, which owned the plant then, conducted a safety overhaul that was supposed to last six months. But there were so many problems with the plant it had to remain shut for two and a half years. Then, in the late 1990s, the second plant at the Indian Point station went into decline. The electrical breakers had not been inspected or maintained, and a second small fine ensued. A year and a half later, the power to the control room went dead, a serious matter, of course; the earlier breaker problem had still not been resolved. While the NRC was considering what action to take against the utility, a tube in the reactor's steam generator ruptured, spilling 20,000 gallons of radioactive water into the plant's sump, and the reactor was taken out of service for most of 2000. A large power company, Entergy, a product of consolidation made possible by deregulation, had bought the two plants in the meantime, bringing their nuclear power production up to nine nuclear plants, and celebrated its purchase in September 2001. A month later, four of the seven control-room operating crews at the number two plant failed an annual relicensing exam. Four months later, a security guard pulled a gun on a colleague in an argument over a glass of orange juice and was

fired. (Kolbert 2003) A trivial matter, of course, but it was a sign of persistent management problems at the operating level.

More problems befell the plant (in addition to a comical terrorist exercise that brought condemnation by a U.S. senator and others). A report by a nuclear engineer declared it one of the five worst plants in the United States. The Los Alamos National Laboratory conducted a study for the NRC, concluding that the chances of a meltdown increased by nearly a factor of 100 because the containment sumps were almost certain to be blocked with debris during an accident. If a pipe broke, the escaping water would go into the basement, and the sump pumps are designed to draw it back into the core to keep it cool. But the tests showed that the cooling water would collect so much debris from the damage as it descended that the debris would clog the mesh screens and prevent recirculation. Within twenty-three minutes at one of the reactors, and fourteen minutes at the other, there could be a meltdown. The NRC knew of this problem as far back as September 1996, but the plant will fix it only by March 2007, giving it eleven years of risky operation. Meanwhile, plant officials said, workers would scour the plant frequently looking for loose material that might become debris. However, the Los Alamos study, which estimated a hundred fold increase in the risk of a meltdown because of the condition, was worried not about debris lying around, but the debris that would be created by a ruptured pipe, fire, or other accidents. (St. Clair 2004)

Finally there was the problem of evacuation after an accident. According to the law, local officials have to sign off on a power company's evacuation plan. This sensible requirement killed the attempt to start the Shoreham nuclear plant on Long Island. Evacuating that densely populated strip of land linked to New York City would have been impossible. (Clarke and Perrow 1996). Massachusetts was unsuccessful in blocking the start of the Seabrook, New Hampshire, plant, right on its border, when it claimed warnings would be insufficient and evacuation difficult. In the case of Indian Point, local officials had, in the past, declared the plan feasible. But after 9/11, new plans had to be approved.

In September 2002, New York governor George Pataki commis-

sioned an independent panel, headed by James Witt, whom we met as the head of a revitalized FEMA during the Clinton administration. It studied the problems thoroughly and could not recommend that the population at risk could be evacuated in time, and said further that it was not possible to fix the plan. This has presented a dilemma for the governor, for New York senator Hillary Clinton (the power company, Entergy, is based in former President Bill Clinton's home state and has been a handsome contributor to Bill Clinton's past campaigns), and for the NRC, and was still unresolved in 2004. (St. Clair 2004)

One of the five worst nuclear plants in the nation, plagued by near misses, and not planning to fix the emergency cooling system until 2007, with no possibility of evacuation for the hundreds of thousands who live nearby and are in immediate danger of losing their lives, or the millions in the New York City metropolitan area that could be contaminated (the nation's biggest "backyard" of "not in my backyard" fame), Indian Point illustrates our vulnerability to industrial disasters, not to mention a category four or five hurricane or, as we have seen, terrorism. We don't have the details, but this would appear to be the case of executive failure as well as the more prosaic management failure.

### We Almost Lost Toledo

Our next story is about the failure of top management at utilities and at the NRC to avoid skating perilously close to disaster. The Davis-Besse plant near Toledo, Ohio, was dubbed by the press as the "reactor with a hole in its head." As a result of unattended corrosion over several years, only a half-inch of stainless steel, instead of seventy pounds of carbon steel, was found to be preventing a nuclear meltdown. But the near meltdown of the plant may also be related to Congressional deregulation and campaign financing and economic power. (Much of this is based on the excellent articles by John Mangels and John Funk, of the *Cleveland Plain Dealer,* an example of local reporting by the print media that kept up the pres-

sure on the utility and the regulators.) The plant is owned by First-Energy Corporation, based in Akron, Ohio. Though there is no direct connection asserted between the plant's operation and the political connections of FirstEnergy, it is worth noting that FirstEnergy is the fourth-largest investor-owned utility in the country, and led the other twenty-nine utilities in donations to the campaigns of George W. Bush. The utilities and the Edison Electric Institute, the industry's technical representative, together raised $6.5 million from 1999 to March 2004 for Mr. Bush or the Republican National Committee. The chief executive officer of FirstEnergy was on Bush's controversial energy transition team, after being one of Bush's fund-raising stars. This is the committee whose minutes the press, environmental groups, and congressional members have unsuccessfully tried to make public. (Henry 2004)

In the 1990s, the NRC issued warnings about corrosion in the penetrations of the reactor head (called nozzles) in boiling water reactors (BWRs). In 1998, the required videos taken of the head of the reactor at the Davis-Besse plant showed significant corrosion, but neither the NRC nor the plant did anything about them. In July 1999, Andrew Siemaszko joined the plant as the lead system engineer. During an outage in 2000, boric acid deposits were found on the reactor head and, according to his account, he attempted to clean them off, but could not finish in one day. He was to return to the job the next day but found all the scaffolding and equipment had been removed, and management had signed a report saying that the reactor head had been fully cleaned. He protested to no avail that a substantial part of head remained unexamined; the company cited the costs of further cleaning and said they could wait until the next scheduled outage. At that outage, in 2002, he was finally able to complete a full cleaning of the head, and discovered a pineapple-sized hole in the top of the reactor vessel created by the boric acid. Subsequently it was determined that acid had eaten away seventy pounds of carbon steel in the vessel, leaving only one-half inch of buckled stainless steel to protect Toledo from a nuclear catastrophe. The Government Accountability Office was to declare it the largest near miss since the Three Mile Island accident in 1979.

(GAO 2004c) Reporting his discovery, documented with videos, he was immediately transferred to another assignment.

There Siemaszko found another problem, this one with leaking reactor coolant pumps. By his account, he persisted in attempts to replace cracked shafts in two of the four that had not been repaired, supported in these efforts by company engineers and an outside expert. Failing to convince the utility that the cracked shafts had to be replaced, he then refused to sign a report saying the issue had been resolved. He was told to sign the report, resign, or be terminated. He chose the last, filed a complaint, and raised his concerns with NRC officials, who, curiously, said his allegations should be handled by FirstEnergy. FirstEnergy denied the allegations and said that he had been fired because poor job performance. He lost a generous salary, excellent health and life insurance benefits, matching retirement contributions, bonuses, paid vacation, and other benefits. In 2005, he was sued by the NRC for failing to report the defects with the nozzle and faces five years in prison and a $250,000 fine, and the company faces a $28 million fine. (Cable 2006) He is fighting the charges. (Funk and Tobin 2006)

The NRC was not unaware of the danger of leaking nozzles that turned up in the Davis-Besse plant. In 2000, before the problems became evident, the NRC had found cooling water leaks from control rod drive mechanism nozzles in many pressurized water reactors. This nozzle sits on top of the vessel and is involved in raising and lowering the fuel rods to decrease or increase the power. Every pressurized water reactor, except Davis-Besse, had inspected the nozzles; many found safety problems with the nozzles and corrected them. The NRC had inspectors who were apprised of the amount of corrosion and the lava-like flows from the top of the Davis-Besse reactor but either did not understand their significance or felt they were not serious, the GAO report noted (with no outburst of incredulity apparent in their report, though one was implied). Worse yet, videos from explorations in 1998 and 2000 had been given to the NRC, showing considerable corrosion. It returned them, presumably without viewing them, despite the several warnings of these problems they themselves had issued. (Mangels and Funk 2004)

In 2001, an alarming case of corrosion was discovered at an-other nuclear plant, and the NRC ordered that all the vulnerable plants be required to either shut down for inspection before the end of the year or show evidence that full inspections had been recently made. All but Davis-Besse complied. Davis-Besse was now in the hot seat, and was ordered to shut down. According to regulations a plant could only operate for six *hours* under these conditions, but the plant argued that it could safely wait until the next refueling outage, scheduled for several months hence. A compromise with the NRC was struck, and the scheduled outage would be moved up and occur in four months, and the plant said it would dedicate one worker to the task of turning on a safety system if the nozzle failed.

When the work was finally done they found the pineapple-sized hole in the reactor's head. As noted, the leaks had been going on so long it left only a thin layer of stainless steel to contain the cooling water in the reactor. The steel was bulging from the severe pressure but had not yet broken. Furthermore, a break would have disabled the safety system the worker was to turn on—the safety program Davis-Besse had proposed in order to keep the plant running. There was so much damage that the emergency sump pump screen would have been blocked by debris, rendering the safety system that was to be turned on inoperable, and the core could have melted. No one should have been surprised; the NRC had issued eleven separate warnings about this danger at plants of this design since September 1988. (Lochbaum 2004; GAO 2004c; Siemaszko 2003)

A subsequent report by the NRC's inspector general office charged that the NRC "appears to have informally established an unreasonably high burden of requiring" of itself "absolute proof of a safety problem, versus lack of reasonable assurance of main-taining public health and safety." (Wald 2003c) In 2004, the GAO issued a blistering report on the NRC's own performance at Davis-Besse, one that rivals the criticisms of the utility that were issued by other government and environmental groups. (GAO 2004c) The GAO's criticisms of the NRC are more significant than those it made of the utility since the NRC is responsible for more than a hundred plants.

The NRC's benign approach to catastrophic risk was fairly recent. It had been somewhat tougher in the past. The change may have stemmed from the deregulation of the electric power industry that commenced in the late 1980s. Shutdown orders were common in the 1970s and early 1980s. As we saw, plants often failed within months of starting up, and the complexity of the new technology meant unforeseen problems in the first decade or so of operation. That shutdown orders are now rare may be due to seasoning of the industry, but near misses began rising in the late 1990s. In the 1990s, the period of rapid deregulation in the energy industry, the commission adopted a policy called "risk-informed regulation." It was designed to pay more attention to the costs it imposed on the plant operators, in order to balance it with the risk reduction that such things as shutdowns afforded. (Wald 2003c)

Documents that reporter Stephen Koff obtained from the watchdog group Greenpeace "show that a special Nuclear Regulatory Commission task force last year [2002] had in fact intended to blame the new regulatory system in part for the slipshod inspections at Davis-Besse. Before the task force's report was complete, however, NRC staff had removed a section on the shortcomings of the NRC's new reactor oversight process. . . . The NRC thus avoided the public criticism that most likely would have resulted if it had more clearly linked the Davis-Besse failure to weaknesses in its new regulatory regimen." (Koff 2003) But the head of the new regulatory regime was promoted two years after his decision to allow the plant to operate without a full inspection of the suspected damage. His new position, the NRC's chief administrative officer, was one step below that of executive director. (Mangels 2003)

The NRC blamed Davis-Besse for not accepting accountability for safety and for not having a "safety culture." The NRC installed an evaluation team in the plant to examine the safety culture and hired outside consultants on safety issues. One consultant was discouraged to find that "senior management has not acknowledged their accountability and responsibility" for the reactor corrosion. Workers, he said, "expressed disappointment and frustration that

this has not taken place." (Wald 2003b) The fired Mr. Siemaszko would presumably agree. He would also agree with the more fundamental criticisms of the regulatory agency for relaxing requirements that encouraged FirstEnergy in its risky practices.

In March 2004, the NRC announced that the plant would be allowed to reopen, which was good news that FirstEnergy needed, since it was facing possible lawsuits from its key role in the massive blackout of Northeast the previous August. (Wald 2004) In chapter 7 we will argue that this, too, was related to deregulation and industry concentration.

As noted, we do not have the evidence to show conclusively that top management was aware of the dangerous condition of the plant and blocked remedial action. Reports by a watchdog group strongly suggests this. (Cable 2006; Smith 2005) But we do know that Davis-Besse came very close to a disastrous accident; a thin, buckled sheet of steel was the only thing that prevented a meltdown. The danger was seen by an engineer who was then reassigned, and when he spotted another danger and refused to ignore it, was fired (though the NRC is disputing this, and blaming the engineer). The NRC ignored the evidence of corrosion in the reactor's head, and when the problem was forcefully brought to its attention again, allowed the plant to run for four more months under dangerous conditions. How many more plants are out there skirting disaster, merely warned by the NRC to develop a "safety culture"? At least one more, as we shall now see.

### Inside the Executive Office of the Millstone Plant

Environmental and public interest groups with highly qualified technical people, such as Greenpeace and the Union of Concerned Scientists, have always been critical of the Nuclear Regulatory Commission for not being tough enough on nuclear plants. But according to a lengthy analysis by John Mangels, a science writer for the *Cleveland Plain Dealer*, the NRC *has* tried at times, and at times it has succeeded. Of course, we expect it to regulate success-

fully, but we should not be surprised by its failures. It has a difficult regulatory problem. The commission is financed by government fees imposed on nuclear utilities rather than from tax dollars. The utilities have a double incentive to minimize the intrusiveness and completeness of the regulations: they can avoid costly changes, and the tax they pay the government will not rise to pay for more intensive regulation by the NRC. Industry's leverage in this respect is through Congress, which sets the taxes that support the NRC and oversees the regulatory process. As we have seen, and will continue to see, unless there is public outrage, Congress is unlikely to demand much of industry and may even demand less regulation.

In the 1980s, largely on its own, the NRC became increasingly concerned about "design-basis" issues, to which it admittedly had paid less attention. These issues involve whether the initial design of the incredibly complex plants met safety considerations. There was simply not a great deal of experience with plants, which, unlike those in Europe and Japan, were of different designs and built by different companies, and even plants built by the same contractor were unique. With increasing experience, the NRC found evidence that plant designs were faulty, that changes in plant hardware and procedures could violate original design requirements, and that records were not being kept about design-based problems. A close call at the Davis-Besse plant, whose troubles we discussed above, in 1985 raised questions about operator behavior, but also about the design of the safety system. In that accident, the plant's main and backup water supplies to the steam generators failed, sending reactor pressure and temperature to dangerous heights and risking damage to the radioactive core. (There were many such management and operations failures in nuclear plants before 1985, of course, making the whole enterprise risky and suspect. For a brief and frightening review of these prior to 1983, see chapter 2 of my *Normal Accidents* [Perrow 1999].)

The NRC cracked down, finding many design-based issues and also a lack of record keeping about design information. It asked the nuclear industry's watchdog group, the Institute of Nuclear Power Operators (INPO), established after the 1979 Three Mile Is-

land accident, to mount an effort to get the plant operators to collect, reexamine, and revalidate their design information. The group declined, saying it was unnecessary since most of their members were doing so voluntarily. Since it was clear that many were not doing so, the NRC pressed on, pledging that it would not fine operators for lesser design deficiencies as long as it was the utilities, and not the NRC, that found them. However, its design-oriented inspections were not just intrusive, but very costly for the plants, and pressure to limit them eventually was successful, as we shall see. (Funk and Mangels 2003)

One of the utilities most concerned about aggressive action by the NRC was Northeast Utilities (NU). Its three nuclear plants in the Millstone facility near New London, Connecticut, had been high performers with good safety records and few unplanned outages (emergency shutdowns, in contrast to the shutdowns for refueling every twelve to eighteen months). But in the mid-1980s, NU's executives called in the McKinsey consulting firm because they feared the effect of federal deregulation on their profits. Profits would be affected by two aspects of deregulation. First, many industrial plants that generated heat during their operations could use the waste heat to generate electric power. If more power was generated than the plant needed, it would be wasted. With deregulation came a new regulation: the utility that ran the power grid had to purchase the excess power of these industrial plants and put it on the grid, paying the price that would be incurred if the utility had generated the power rather than the industrial plant. This would increase supply, and that would lead to lower prices. It was quite sensible, but not expected to amount to much increased supply for several years, and it didn't.

The second effect of deregulation that concerned NU was the regulation that any utility could have access to any other utility's transmission lines. NU's grid would be open to other wholesaler power producers through a bulk regional power pool. If a utility in New York State offered cheaper power than NU in Connecticut, NU's large customers could have it sent over NU's lines and purchase it. NU no longer would have a monopoly with power generation in

the region assigned to it. This also would lead to lower prices, though this regulation was expected to have little effect at first.

(In what follows I am following the extremely detailed and well-documented account of NU's troubles by economists Paul MacAvoy and Jean Rosenthal [2004]. As a result of lawsuits, voluminous company records became available for their research. Unfortunately, there is no comparable inquiry into other nuclear facilities, such as those at Davis-Besse or Indian Point, since no other utilities have had their records exposed so extensively as Northeast Utilities had. This bonanza offers us the only glimpse we have of what takes place in the executive suite of nuclear power plants.)

Utilities throughout the nation formulated strategies to deal with the new competition; they anticipated that prices would fall and they could lose money. The dominant strategy, more talked about than followed, was to diversify. Awash in cash, the utilities could afford diversification. (MacAvoy and Rosenthal 2004, 23–24) But most utilities and industry authorities thought that falling prices would be in the distant future. NU was highly profitable, but it said it expected intense price competition soon, by 1990. Along with FirstEnergy in Ohio, NU's management and its board of trustees decided to meet the competitive threat by cutting its operating costs, most particularly those in its three nuclear power plants, where the operating costs considerably exceeded that of its oil and coal-fired plants. Even more specifically, it would reduce operating costs not by improving operating efficiency, as some nuclear plants successfully did, but by reducing *maintenance costs* and *employee costs* at these plants. Their management consultants at McKinsey did the justifying study.

MacAvoy and Rosenthal write that NU "took risks with plant operating rates in order to take deep cuts in current maintenance expenses."(27) (It is disturbing that the two economists defined the risks only in terms of shareholder value, with scant acknowledgment that there would be risks to the laid-off employees and, more important, should the cuts cause a major accident, to the general public near the plant! Economists call such risks "externalities" and rarely consider them.) The top management knew the risks of

the cuts. The authors quote an internal memo that notes the consequences of the cuts clearly. While the strategy had produced unprecedented profits by 1990, the memo acknowledged that the plants had excessive overtime, were losing qualified personnel, and were unable to train instructors, to conduct required inspections, or even to do adequate safety reports. (45) As a result, forced outages occurred for extended periods in 1990 and 1991—for example, those due to corrosion in pipes they had failed to maintain—and then they failed to identify the corrosion.

Forced shutdowns are expensive; power must then be purchased elsewhere or generated from more expensive emergency internal sources, and repairs are expensive and may take months. For this reason, economists and engineers regularly reason that firms have an interest in preventing accidents and will do the necessary maintenance and have the necessary workforce. Lewis and Darken are typical. They say: "Continuity of operations already has its own built-in motive—the more reliable the operation, the more money received. Therefore, utility companies are motivated to increase continuity of operations. They do not need governmental incentives to reward them for doing what they do best: deliver services and consumables to the public." (Lewis and Darken 2005)

But the risk of incurring forced shutdowns is considerably reduced in the short run of a year or two or more by the "regulatory accounting" practices of the NRC. This limits the impact of the shutdowns on earnings, since *the expenses are deferred*; they are expected to be covered by anticipated future revenues. (61) The cost of the shutdown does not reduce profits because it is charged to the future in anticipation of profits then. In effect these NRC accommodations reduce the incentive to avoid accidents, which is the opposite of what the NRC should be doing.

Alarmed by the forced outages, the NRC augmented its full-time inspection team at Millstone. It also determined that employee morale there was poor, employees were harassed for reporting safety violations, and experienced employees were leaving. The nuclear facilities operated less than half the time in 1991. (48–50) Internal task force reports prepared by the company itself were dev-

astating. One argued that engineering should attain "excellence in performance and not cost containment," a direct rebuke of the NU/McKinsey strategy announced in 1986. (53) The NRC was quite aware of what was going on. A special review group report by the NRC charged the company with micromanagement (an aseptic term for telling workers to violate requirements), harassment of employees, and an overemphasis on cost containment. (56) Later, a NRC official worried that if NU succeeded in buying the bankrupt nuclear plant at Seabrook, New Hampshire, the "Millstone virus" (i.e., mismanagement) might spread there. (60) (The cost containment strategy was identified as a "management failure" by the NRC, considering it rather like a poor decision or poor strategy rather than a deliberate act.) Incredibly enough, with this poor record for its nuclear plants, the NRC allowed NU to purchase Seabrook, under the vague conditions that it improve its employee relations and other safety matters at Millstone. But improving conditions at Millstone would mean abandoning cost containment and short-term profit goals by the company, so conditions failed to improve. MacAvoy and Rosenthal dryly note that the commitment to safety would itself "be subject to cost containment later in the decade." (61)

A stock analyst or potential investor would have no reason to be wary. That fallible index of company health, its profits, increased each year after 1989, when the program was in full swing. The cost containment program continued to generate higher profits. Cost containments in the nuclear facilities realized savings of nearly $70 million a year; payroll savings in 1991 were estimated at $27 million annually since 1987; large workforce reductions of 1,100 were achieved. (62) Maintaining this strategy, MacAvoy and Rosenthal write, "would add to earnings, adding to management's own salaries, bonuses and stock awards, with expectations of continuation well into the 1990s." (63) The competition expected in 1986 by the company and McKinsey had yet to arrive.

Where was the regulatory agency in all this, as the plant cut staff and operating costs, neglected safety, angered its employees, and increased profits and management bonuses and stock options? The

NRC *was* busy; it established a full-time oversight group at the Millstone site, increased the number of inspectors, required the utility to respond to a list of requirements, increased the number of fines, and discussed the utility's performance at eleven NRC senior management meetings between June 1991 and 1993. (74) That should make a difference if the problem was either management failure or operations failure, but it didn't. Why was the NRC not successful? Because it identified the problem as a management problem that would be responsive to lists of requirements, oversight, and fines. MacAvoy and Rosenthal are not taken in by this analysis, as plausible as it may seem. The problem, they say, was the executive strategy adopted, not the failure of management, and this strategy was adopted because of deregulation. An essay by a well-known energy stock analyst in NU's 1991 annual report suggests how this could happen: touting American capitalism, he said the electric energy industry "is not a regulated industry any more, it is a dynamic, market-driven enterprise." (65) With the deregulation this entailed, NU could ignore the NRC's requirements lists and pay its paltry fines.

In May 1993, the NRC chastised the firm "for the harassment and intimidation of a Northeast nuclear supervisor who had raised safety issues." (75) NU paid a small $100,000 fine, but operating earnings were up 15 percent that year. A new violation took place from May to August of 1993. Instead of shutting down the plant for two weeks to repair a vital leaking valve, they drilled into it to inject sealant into the gasket, and then kept striking the spot with a pneumatic ball-peen hammer to close the hole. This crude method failed, since the hole would open again. They attempted to close leaks by striking the spot thirty times over the next months, until this caused one of the studs supporting the valve to fail, causing the reactor to depressurize and go into a forced shutdown. The stud breakage, the NRC concluded, risked a "loss of coolant accident" (LOCA) and potential meltdown. (76) This was serious. Since the "errant repair activities" had continued for weeks, one wonders what the full-time NRC inspection team was doing, along with the regular NRC inspectors?

There were other inspectors, hired from a contracting firm, in the plant, and they observed the repair attempts. But they "worried that bringing up safety issues might affect their future employment." One of these inspectors, who thought there was a "significant safety risk," still signed off on an inspection of the valve "to prevent a confrontation with management." (77) One wonders why are they called inspectors, but recall that FirstEnergy fired its chief engineer for refusing to sign off on a faulty valve at the Davis-Besse plant.

The NRC dutifully lambasted the utility (but apparently not its own inspectors) for all sorts of weaknesses and failures. The next year (1994) the two top executives at NU realized 37 percent increases in their compensation packages. (84) In 1995, the industries' "self-regulator," the Institute of Nuclear Power Operators—which is funded by the industry and conducts serious investigations but makes none of its findings public—met with the board of trustees of NU, but there is no documented board response to this unusual event. Also unusual was a 1995 meeting of the executive director of the NRC and his staff with the NU board. The NRC participants in the meeting laid out the politely labeled "lingering performance problems" in considerable detail, and the trustees promised to do better and "expressed appreciation for the meeting." (80) But NU, in the next year, got its board of trustees' approval for an even more ambitious cost-cutting program, 40 percent over five years. (78) The failure of regulation (or should we say the success of deregulation?) and the profit-maximizing strategy of the plant's executives continued to put the plant on the brink.

### Enter the Whistle-blower and the Media

In 1996, one of those dramatic stories about a near miss at a nuclear plant made the cover of *Time* magazine, the first cover story of an industry in the magazine's history. It revealed that the Millstone nuclear plant had operated for twenty years "outside its design base" (violating the operating conditions required by its basic design) by routinely unloading all the spent fuel rods from the re-

actor core at once, during refueling operations, rather than one-third at a time, as regulations required. It also was doing so without waiting the required time for the radioactive fuel bundles to cool. The unloading was so rapid that it melted workers shoes at times. The cooling system was stressed with so many hot fuel rods being put in the pool at once. (The cooling system was subsequently discovered to be inadequate even with partial insertions, and especially inadequate for what was going on: receiving the contents of a complete unloading of the fuel rods being removed from the core.)

The *Time* account detailed the stubborn efforts of one employee, George Galatis (with the initially reluctant support of another), who discovered that the routine practice was outside of the design-based limits—that is, should not have been happening. The employee took the case to the Millstone management, which refused to do anything about it. After eighteen months of pressing the safety case, he took it to the NRC itself, only to find that the commission knew about the unsafe practice but had made no move to stop it.

We have to turn to journalistic accounts here; the incredibly detailed work of MacAvoy and Rosenthal is focused on shareholder value, not on the regulation of a plant with catastrophic potential, so it omits many of the following details, though it cites the *Time* report. The NRC, according to *Time,* said the practice was common and safe, if the plant's cooling system is designed to handle the heat load, despite explicit regulations to the contrary. By design-based criteria it was unsafe, but the NRC would ignore that. However, the pool and its cooling system had never been examined to see if it could handle a full dump, and that is what it had been getting. (Pooley 1996)

The outcry occasioned by a dramatic cover story energized the NRC. Its new head commissioner, Shirley Jackson, began a crackdown, scaling back a policy called "enforcement discretion," which had allowed the agency to set aside hundreds of its own safety regulations. Millstone had received fifteen such waivers since 1990, though none for dumping. The NRC inspector general, who in-

vestigates agency wrongdoing but has no power to punish, said, "We shouldn't have regulations on the books and then ignore or wink at them." (Pooley 1996) Commissioner Jackson said that with only four inspectors for every three plants they could miss things (but this was a big thing to miss, and they had put on extra inspectors). That would mean fewer than fifty inspectors out of three thousand NRC employees. As one commissioner said defensively in the 1980s, they were not an inspecting organization, but an accounting one—they mainly keep records. But, Commissioner Jackson of course said that having this few inspectors never endangered safety. The watchdog environmental agencies were outraged.

Fortunately, MacAvoy and Rosenthal do give us a lead as to why a plant like Millstone (or the other two we examined), with so many violations and fifteen waivers in six years, could have inspectors that would miss something that several employees were worried about and one of whom had filed a safety report with management concerning dumping. The NRC requested that its Office of the Inspector General investigate the dumping incident. "The investigation determined that NRC resident inspectors at the Millstone site were aware of the practice of full-core offloading, but *did not know the design basis well enough* to realize that the practice was contrary to the facility's license. The investigation also determined that there had been no analysis of the heat removal capacity of the Millstone One spent-fuel pool cooling system under conditions of a full-core offload." (MacAvoy and Rosenthal 2004, 89; italics supplied) Since failure of the storage pool would be catastrophic—widespread radiation would require immediate evacuation of the plant, perhaps kill employees, and leave the nuclear reactor unattended, and could contaminate the surrounding area while killing thousands—one would think that an inspector would inquire into the design basis of such a vital element. But, the *Time* report charges, it is even worse. The NRC home office was aware of the illegal Millstone practice and even countenanced it at other plants. Perhaps this is why nuclear engineer Galatis resigned his position in 1997. The country lost a nuclear engineer (it is not a well-stocked profession) because he entered divinity school.

Another whistle-blower was not so fortunate. The NRC, though chastising the utility for threatening whistle-blowers in the Millstone case, retaliated against one in another case, as the *Time* magazine story reports in its cover story. In the early 1980s, when Northeast Utilities' Seabrook Station in New Hampshire was under construction, Joseph Wampler warned the NRC that many welds were faulty. His complaints went unanswered, and he was eventually fired. He moved to California and sought employment in nuclear power stations. But in 1991 the NRC sent a letter summarizing Wampler's allegations—and providing his full name and *new address*—to several dozen nuclear companies. His career was destroyed a second time; he now works as a carpenter. The NRC fined NU $100,000 for problems with the welds. (Pooley 1996) At least the NRC has not tried to jail Mr. Wampler, as they are trying to jail Mr. Siemaszko.

It is apparent that we cannot depend on whistle-blowers to protect us from the NRC or the nuclear power plants. They are fired, and they have trouble suing if they try. And even if they don't sue, they can be blacklisted or jailed by our protectors, the Nuclear Regulatory Commission.

The utility stumbled on with more problems until finally all three nuclear plants were shut down and the NRC would not let them restart without extensive reorganization, refitting, training, and so on. Millstone One had not been allowed to restart after its late 1995 refueling. Millstone Two had been down for ten months after having restart problems after a 1994 refueling, restarted in August 1995 but failed in December 1995, restarted briefly in February 1996 but had to be shut because of "operating problems," and underwent extensive review. Millstone Three had even more problems. The four pages that MacAvoy and Rosenthal devote to a detailed analysis of the faults of the company and the three reactors in this short time period is depressing reading, especially since there had been so many warnings and futile efforts by the NRC to secure reforms. (90–94)

One of the themes of redemption in the Millstone saga was changing the "safety culture." The NRC made a valiant effort;

MIT researchers made studies; consultants were hired; and, as happened with the NASA *Columbia* shuttle disaster years later, a failed safety culture was said to be the root cause of the failures. After a long history of breakdowns and violations, the NRC addressed the issue of organizational culture. As summarized in a *New York Times* story, the NRC "will not allow the reactors to restart until management completely changes its culture and proves that workers feel comfortable when they raise warnings about nuclear safety." Plant operation had been so shoddy that managers routinely broke federal rules. "When workers raised concerns about cooling pipes wrapped in duct tape, faulty gauges, torn filters and the mishandling of fuel rods, they were not only ignored, they were punished." (Rabinovitz 1998) A huge effort over the next two years to change the culture followed.

But was this the root problem? Safety cultures can only be developed where top management wants them, or at least will tolerate them. (Those of us who have worked with NASA after the *Columbia* disaster will testify that changing a culture is extremely difficult when "efficiency," "privatization," and "cost reductions" remain the top management's primary concern; their culture can trump all others.) The limits to this approach, changing the organization by changing employees' and operating managements' culture, are evident in the Millstone case. A safety culture was not in the economic interests of top management, so it could not take root. (For a contrary view on the efficacy of safety cultures in high-risk organizations, see the literature on high reliability theory [La Porte and Consolini 1991; Roberts 1993], and for its application to nuclear power plants, see the fascinating ethnographic account of Constance Perin [2006].)

### The Collapse of "Shareholder Value"

The Connecticut Department of Public Utility Control (DPUC) is the hero of this case. It held public hearings and brought suit against Northeast Utilities. This shows what persistent regulatory inquiry can achieve despite the NRC. Also helping was a lawsuit

brought by ten small utilities that had contracts to buy power from the company. (MacAvoy and Rosenthal provided expert testimony on the behalf of these utilities, which gave them access to data that researchers normally would not receive.) The public hearings of the DPUC, the data that the NRC was forced to disclose, and the disclosures from the lawsuits provided a rich treasure trove that gave the authors this unprecedented access to executive and board actions.

The Connecticut public utilities authority ruled that NU could not make its customers—households and other consumers—pay for the cost of the shutdowns, declaring that the company was mismanaged and its executives were responsible. In 1999, NU pleaded guilty to twenty-five felony violations of environmental and safety regulations at its power plants between 1994 and 1996, paid a $10 million fine, and declared bankruptcy. The NRC finally allowed two of its reactors to restart, and all three were sold to Dominion Resources of Virginia in 2000. The company sold its fossil fuel plants and Seabrook, and now is only a common carrier, distributing power others produce. (MacAvoy and Rosenthal 2004, 4, 106)

The authors found it depressing that the board of NU failed to oust the head of the company, Bernard Fox, after a disastrous financial collapse. They noted a "stunning" lack of urgency on the part of the board to do anything about the failed leadership. When Fox finally retired later in 1997, he left with a handsome financial package, even though the company was hemorrhaging financially and threatened by lawsuits. The handsome payments to the two top officials upon retirement "were free of penalties against senior management for destruction of the company," MacAvoy and Rosenthal noted in their understated, bleak style. (108) A litany of organizational failures was acknowledged by a new president of nuclear operations, but as the authors sagely noted, "this litany addressed management process, not the faults in key strategy decisions." (101–5)

This observation is very important. Enterprises with catastrophic potential are vulnerable to organizational failures, such as bad management and failures at the operating level. But MacAvoy and

Rosenthal are raising a more fundamental point, though they do not make it as explicit as I now will. The management failures so often detailed in NRC reports and those of consultants and the Connecticut Department of Public Utility Control—threatening employees, falsifying reports, failing to make mandated changes, cutting maintenance, reducing the workforce unreasonably, providing poor or no training for employees, and so on—were not inadvertent or the result of lack of knowledge, experience, or ability or of overwork or time pressures (the usual causes of poor management) but were *intentional,* the result of *conscious decisions.* Top management, with the knowledge and passive support of the board of trustees, knowingly *caused* these practices.

The "management failures" were an inescapable and very visible consequence of NU's decision to increase profits by cutting costs. (We do not have the data on Davis-Besse and FirstEnergy to support similar charges, but the cases appear to be very similar.) The managers expected that deregulation would bring about competition that they had not experienced before; in the past, with no competition, their rate proposals were accepted or modified by Connecticut state officials. In anticipation of this competition they chose to meet it by progressively, and ever more drastically, cutting maintenance and operating costs. Under pressure from the NRC, they promised changes but demonstrably never made them; indeed, they prevented their managers from carrying them out.

*Management at the organizational and the operating level did not fail; they obeyed.*

This entailed a risk. Senior managers ran the risk of having forced outages and even of receiving NRC orders to shut down or not restart (as well as the risk of the collapse of a radioactive core, considerably more dangerous than the "collapse of shareholder value"). For about a decade the risk paid off; profits went up and the company expanded to include the Seabrook reactor. Then the risk stopped paying off. With all three of its reactors closed and the company being sued, and profits turning into losses, the two top managers "retired." Despite the restart of two of the three Millstone reactors, the company was "destroyed," that is, forced to sell off its production facilities.

We will see many examples of organizational failures where poor management is to blame; management is difficult, and there will be cases where it fails. But that must be distinguished from what MacAvoy and Rosenthal call "strategic failure," where top management (and the board) knowingly takes risks in violation of regulatory laws. I prefer the term *executive failure* to strategic failure, and I will use this term in the rest of the book.

Executive decisions commit the whole organization; decisions by organizational members below the executive level do not do so, though sometimes inadvertently they will. A particular department may make a decision that maximizes its benefits within the organization, and this may weaken the organization as a whole, but this is not the intent. An executive decision to maximize the executive's interest in his bonus or stock portfolio obviously commits the whole organization; it is hard to argue that this was inadvertently contrary to the organization's interests. The executive should have only the whole organization's interests at heart; that is what it means to be the head of the organization.

In contrast to executive failure (knowingly harming the company the executives are responsible for in the face of outside and inside warnings), strategic failure, MacAvoy's and Rosenthal's term, means a faulty strategy was selected without awareness of its consequences. If McKinsey had recommended a strategy that some other companies had used to meet the projected competition, such as increasing the efficiency of their plants and spending their ample capital reserves on diversification, that would be a strategic success if the executives adopted it. But something more was operating here, and for those of us interested in the safety of our population, it is more important than these strategic choices. The top executives were ordering their subordinates to break the law and to take risks their subordinates—managers and workers—knew to be unwarranted. Thus, the *executives* failed, not just their strategy.

"Our general interest is not the technology of safety," the authors serenely declare, but how the trade-off of profits versus safety "may be part of corporate strategy." Management, they continue, "for its own advantage, carried out a strategy that was too risky to benefit the corporation," that is, the financial stakeholders. Man-

agement was acting against the interest of investors, "who would not have placed such a high priority on current earnings and executive compensation." (xi) (But they provide no evidence that investors did not view current earning as a high priority; one might assume investors always put earnings as a high priority.) "The company took on the risk of destruction, while the decision-makers were left relatively unscathed," they conclude in their final sentence. (111) Reputations were damaged, but the two top executives, Bernard Fox and R. W. Busch, benefited financially, receiving $1.6 and $3.0 million in voluntary separation settlements. (108) The nation was to see much more of this in the corporate scandals of 2000–2004 and beyond, but those risks did not involve the risk of millions of deaths and environmental destruction inherent in our nuclear power plants.

### THE NRC AND CONGRESS – CAPTURED?

More was involved than executive failure by officers and top executives of the company, and we have to go beyond our economists to see it. The failures had to be allowed by our government for the executives to carry out ten years of destruction before it was called to a halt. A year after its cover story, *Time* revisited the case, and found that NU had made the fuel-pool cooling-system changes that employee Galatis had demanded for eighteen months. The NRC only then admitted there was "pervasive noncompliance" that did pose a potential threat to public safety. (Pooley 1997) Perhaps it was the NRC that had been a threat to public safety. How did it allow NU to go on, year after year, pursuing profits and increases in top executive compensation packages at the obvious expense of safety? We probably have to give more thanks to Connecticut's Department of Public Utilities Control, which badgered the utility for years and sponsored slashing reports by consultants, for bringing NU to its heels, than to the NRC. Connecticut's actions helped bring about the ignominious demise of the utility. It brought suits.

As we saw with other examples of regulatory failure, there are more actors in the picture than just the regulatory agency. There are the industry trade associations and, most particularly, the congresspeople who receive their campaign donations and other favors. Here is what happened when the NRC got tough, as it sometimes does. Writing of the agency's actions in 1997 after the publicity about Millstone, journalists Funk and Mangels observe: "After a decade of talking tough about rooting out unaddressed design flaws and uncertainties system wide . . . the NRC was doing something about it." (Funk and Mangels 2003) In that year, an unprecedented number of thirteen facilities were put on the watch list for extra scrutiny; the more the NRC looked, the more flaws the special inspection teams found; and reactors were shut down. But there was this important consequence of an agency finally doing its job, says Mangels: "operating companies' profits took a nose dive."

When this happened, the NRC slackened its inspections. The agency and the industry made the argument that by the latter half of the 1990s, the plants had run long enough to disclose any serious flaws, and the remaining design-basis issues were not threatening public safety. One wonders how so many violations could be insignificant, and if they are insignificant, why are they called violations? There are many such issues. The antinuclear group Public Citizen identified more than five hundred cases of design-based flaws between 1996 and 1999. A NRC study, with much more access to facilities and more manpower, found a great deal more violations: 569 in one year alone. There were more than 3,100 from 1985 to 1997, an average of about 240 per year per reactor. The chief of the NRC's inspection program branch said of three hundred discrepancies over a year old at the Davis-Besse plant in Ohio, "they're minor items that, even if they're not corrected in a short period of time, it's really not going to contribute to risk." This is a NRC inspector talking. (They were not all minor, as we saw; some threatened meltdowns.)

In 1997, a year after the Millstone disclosures and forced closing, and in the period when the Davis-Besse plant was ignoring signs of damage to the top of the reactor, all very serious matters,

the head of the Institute of Nuclear Power Operators (the private industry "watchdog" research agency) supported the complacent views of the NRC inspector just quoted. The head of the INPO warned that plant managers "are often involved in fairly esoteric issues with the NRC when their time would be better spent focusing on the day-to-day safe and reliable operations of the plant and performance of their people." (Mangels 2003) But the issues the inspectors at Davis-Besse and Millstone One, Two, and Three should have been focusing on were hardly esoteric. They should have been looking at a design-basis danger the NRC had repeatedly warned of in the 1990s.

I am sure the NRC can engage in nit-picking and be involved in esoteric issues that do not affect safety; one reason is that until the accident unfolds we often do not know what is esoteric and trivial and what is not. The streaking from the top of Davis-Besse reactor dome was considered trivial, but, as we saw it, was not. Full core loading in a storage pool not designed for it might have been trivial, and unrelated to safety (though that is hard to believe), but turned out to be so risky, when examined, that even the utility agreed the storage system had to be greatly modified.

The problem with considering the thousands of regulations issued by the NRC as picky and hamstringing is that the regulations are a response to violations. Initially, the predecessor of the NRC, the Atomic Energy Commission, issued few violations, but as the plants came closer to coming on line, it had to step in with regulations. New problems with the behavior of the contractors, then the operating utilities, engendered more regulations. The variety of designs and the uniqueness of the plants required still more. The multitude of regulations (many became irrelevant, as designs and practices changed) was a response to a multitude of unsafe practices by the utilities. Violations at one plant produced stricter regulations at all similar plants—for example, the boiling water systems and the pressurized water reactor systems, the two major styles. It is similar to civil laws and criminal laws: they are enacted to deter undesirable behavior, not fashioned out of thin air.

With utilities profits falling when the NRC got tough after the *Time* story, the industry not only argued that excessive regulation

was the problem, it did something about what it perceived as harassment. The industry used the Senate subcommittee that controls the agency's budget, headed by a pro-nuclear Republican senator from New Mexico, Pete Domenici. Using the committee's funds, he commissioned a special study by a consulting group that was used by the nuclear industry. It recommended cutting back on the agency's budget and size.

Using the consultant's report, Domenici "declared that the NRC could get by just fine with a $90 million budget cut, 700 fewer employees, and *a greatly reduced inspection effort.*" (italics supplied) The beefed-up inspections ended soon after the threat of budget cuts for the agency. (Mangels 2003) And the possibility for public comment was also curtailed, just for good measure. Public participation in safety issues once was responsible for several important changes in NRC regulations, says David Lochbaum, a nuclear safety engineer with the Union of Concerned Scientists, but in 2004, the NRC, bowed to industry pressure and virtually eliminated public participation. (Lochbaum 2004)

As Lochbaum told reporter Mangels, "The NRC is as good a regulator as Congress permits it to be. Right now, Congress doesn't want a good regulator." (Mangels 2003)

In a remarkable charge, *Time,* in 1997 said, with reference to the NRC, "The industry vetoes commission nominees it deems too hostile (two of five NRC seats are vacant), and agency officials enjoy a revolving door to good jobs at nuclear companies such as Northeast." (Pooley 1997) I have heard such charges from a former NRC regional executive, who left the agency in frustration. The regional executives in particular are said to be informally subject to utility vetoes over appointments. But I know of no evidence that would prove this charge by the magazine.

## CONCLUSION

I have argued that nuclear power plants have unusual safety problems, as compared with the rest of industry, because of the complexity of their designs and operations and their catastrophic po-

tential; that these take time to reveal themselves; and that their aging and the granting of extensions on their operating lifetime pose new risks for which we have no experience. Furthermore, what had once been a fairly comfortable business environment with little or no competition (and undoubtedly uneven pricing and inefficiencies), has, with the deregulation that started in the 1980s and began to take effect in the second half of the 1990s, encouraged a "race to the bottom" in terms of maintenance, inspections, employee staffing and quality, and safety cultures. Not all plants succumbed to the pressures to increase profits in a competitive environment, but enough have to place parts of the nation at serious risk, especially those plants close to major urban areas.

The complexities of these plants makes it particularly hard to prevent operation failures (employee errors, poor teamwork, etc.) and management failures (the organization of work, supplies, facilities, and other resources; the safety culture). Many failures have these roots. In *Normal Accidents* I argued that because of the complexity of these plants and their tight coupling, serious accidents are inevitable *even with the best management practices and attention to safety.* (Perrow 1999) (Nuclear power plants with passive safety systems have been proposed and could be built; I am concerned about the ones we are stuck with.) But the most serious failures are those of the top executives, who knowingly require unsafe practices by their managers and workers in the interests of profit maximization.

These vulnerabilities of nuclear power require a vigorous regulatory effort, especially since there is no meaningful liability penalty for a catastrophic accident. The regulatory effort has been mixed and episodic, with suggestions that the regulators are, as the political science literature puts it, "captured by the industry" they are supposed to regulate. We have seen evidence of inspectors who do not inspect, or even understand the system as well as first-line supervisors and workers, of a huge bureaucracy that is committed to bookkeeping rather than on-site inspections and only manages to energize itself after the publicizing of major failures. This suggests a major role for environmental watchdog groups, funded by tiny donations and staffed by dedicated professionals, to watch

over the regulated and the regulators alike. It also suggests the vital role of the media, especially the print media, in covering the major failures.

But behind the regulatory failure lies a more basic one, the power of Congress to threaten the regulatory agency with fiscal punishment if it is aggressive. The energy industry is rich and huge, and its campaign contributions are massive. It also contributes to the infusion of an economic ideology of deregulation, free markets, competition, and efficiency. Actually, competition is declining as markets are consolidated and made less free, and the efficiency is aimed at short-term profits rather than long-term investments in safety and reliability. If industry has captured the regulatory agencies and promoted a free-market ideology, we have Congress and our campaign-financing system to blame. Congress writes the regulations and influences the judiciary that enforces the laws.

Nuclear power plants concentrate more lethal potential than anything else in our society. They are vulnerable to natural disasters. There have been emergency shutdowns in the face of hurricanes, for example, though no storms or floods have as yet disabled a plant's external power supply and its backup power generators. Some plants sit on earthquake faults. This chapter has argued that they are extremely vulnerable to terrorist attacks and to organizational failures. Their electricity is considerably more expensive than alternative means of generation, and while they pollute far less (only in the short run; in the long run of thousands of years, their wastes pollute far more if they are not contained) and release no carbon dioxide, the current difference between oil- and coal-fired plants and nuclear plants in this respect could be greatly reduced if currently available emission reductions were required of fossil fuel plants. And, of course, the federal government invests only a trifling amount in research on solar and wind power and energy conservation, while it continues to handsomely fund nuclear power research. This is an example of increasing our vulnerability to natural, industrial, and terrorist disasters. By supporting pollution reduction from fossil fuel plants, alternative energy sources, and energy conservation, we could phase out our vulnerable nuclear plants in a decade or so.

# 6  Better Vulnerability through Chemistry

------------------------

ON JULY 18, 2001, BALTIMORE was at risk of devastation when a sixty-car train owned by CSX caught fire in a tunnel in downtown Baltimore. Several of the cars had pulp material, which caught fire, burning other cars with dangerous chemicals, one of which was hydrochloric acid. The train derailed at three in the afternoon, and by nine that night all major highways into the city were closed to incoming traffic. Local telephone service was disrupted, a baseball game canceled, and parts of the backbone of the Internet were slowed. A water main broke, flooding and collapsing streets and knocking out electric power to 1,200 people. Even weeks later it was not over; a combustible chemical appeared in city sewers but fortunately was not ignited.

But as we always say, following a disaster, we are lucky it wasn't worse. Had a railcar carrying ninety tons of chlorine (a routine event on the railroads) been one of the cars on that train, an estimated one and a half million people could have been at risk. As a FEMA report put it, "The derailment was unusual; the shipment was not. CSX reports that 40 freight trains run through Baltimore on an average day. Some days, all of them carry hazardous materials." (Clarke 2005, p 117) Among the hazardous materials carried by rail are caskets of high-level nuclear waste from our power

plants, and if the Yucca Mountain or some other facility to store the waste for centuries goes online, there will be daily shipments going through cities and tunnels. We have been reassured that the caskets could withstand extreme heats, up to 1,475 degrees for thirty minutes. But the fire in the Baltimore tunnel (which would handle shipments from Maryland's Calvert Cliffs nuclear power station) exceeded that temperature for a full twenty-four hours. Experts estimated that twenty-four square miles around the accident (Baltimore, neighboring towns, and much more) would be highly contaminated or worse. "Amazingly," say Clarke, whose account in *Worst Cases* we are drawing on, "officials in the Nuclear Regulatory Commission think the Baltimore experience was actually a success story." (118) They said the robust nature of the casket was evident and the exposure in the tunnel fire would not release any radioactivity. The data Clarke reports flatly contradicts our Nuclear Regulatory Commission.

### INTRODUCTION

Really devastating chemical-plant disasters have been rare in the last few decades. There is, of course, Bhopal, which we will come to, but the yearly deaths from individual chemical accidents in the United States is relatively small. The industry kills from 250 to 260 people a year, but this is hardly noticed since it is usually only two or three a time—and they are almost always workers—and most of these are transportation accidents rather than those that occur at industrial facilities. (Johnson 1999) (Think of the outrage if we had comparable fatalities from two 727 jetliner crashes each year, especially since those are victims with higher status than workers and can afford air travel.) Accidents are very frequent, averaging five a day, but so far the devastation has been remarkably small, given the potential. (Purvis and Bauler 2004) Hurricanes have rarely made direct strikes on chemical facilities. None of the chemical facilities accidents to date have resulted in catastrophic emissions, though evacuations are frequent. Even Bhopal, supposedly a

wake-up call for the industry, did not prompt any industry-wide efforts of significance. But the rise of terrorism in the late 1990s prompted a great deal of concern about the security and safety of chemical plants and other concentrations of hazardous chemical materials. The chemical industry took notice and established safety standards, a government board increased its activity, public interest groups sounded alarms, and after 9/11, journalists wrote about walking up to unguarded tanks that could endanger seven million people with one terrorist suitcase bomb. The catastrophic potential of this vital industry is now on the map. (We will not consider the long-term health effects of other aspects of the chemical industry. We are only interested here in large catastrophes.)

The themes of this chapter are familiar:

(1) Organizations are always imperfect tools, and the best of them cannot provide complete security from the determined terrorist, avoid the devastations of weather, or avoid the inevitable mistakes that lead to accidents.

(2) The chemical industry, as with the nuclear power industry and others we will consider, will have a preponderance of firms that do not try very hard to prevent the occasional disaster. Prevention costs money in the short run, and in competitive capitalism, short-run advantages pay off, even if they do not in the long run.

(3) The institutional framework (Congress, regulatory agencies, trade associations, unions, etc.) is not conducive to security, safe operations, or protection from extreme weather. The self-regulation scheme of the chemical industry is ineffective, and our local and national government either is insufficiently aware of the catastrophic potentials, too beholden to the political campaign blandishments of the industry, or, as many organizations will always be, indifferent to the public welfare. All this would change if, instead of a dribble of the equivalent of two 727 crashes each year, we had two Bhopals or two horrendous terrorist attacks a year for, say, five years.

(4) There is an alternative that would not depend on having perfect organizations, and is not inconceivable even with our flawed institutional framework: reduce the size of the concentrations of

hazardous materials, and reduce their toxicity and potential for fires and explosions. A few changes in this direction are being made; they could be extended to many of our 1,500 risky sites, at some economic costs, of course. With smaller targets and less-potent substances, terrorist acts would have less effect; the five-a-day accidents would be even less consequential; and the weather would take a much smaller toll.

(A note on terminology. I will generally cite the danger as "tox-icity," but in strict terms there are four categories: *toxicity,* or the lethality of a chemical to human health; *volatility,* or the tendency to disperse the gas phase; *flammability;* and *reactivity,* or the tendency of a chemical to release a large quantity of energy or explode spontaneously. (Ashford et al. 1993, III-18) The term "hazmats" embraces the lethality of these four, and toxicity will be the general term.)

## BHOPAL, THE MOTHER OF THEM ALL

The largest chemical-plant disaster in history, the release of poisonous gas on December 3, 1984, at the Union Carbide plant in Bhopal, India, had "prompt deaths" of at least 2,000, and perhaps as many as 6,000 to 8,000 prompt deaths, and left another 100,000 injured, many of them permanently. Following the explosion, a toxic cloud of methyl isocyanate drifted over the sleeping citizens in an adjacent shantytown, penetrating their porous homes. No alarms were sounded and when some phoned the plant, officials denied there had been a release, even as most workers were fleeing upwind. The cause was a commonplace, prosaic industrial accident of the kind that will continue to occur. Union Carbide still insists that it was sabotage, thus denying responsibility. Others say the immediate cause was an "operator error" in connecting some hoses, though the source of this explanation was the Arthur D. Little Company, which is a consultant to Union Carbide. A U.S. Chemical Safety and Hazard Investigation Board (CSHIB) report targets the company, rather than a hapless operator, saying that

there was an explosion due to a nonfunctional critical part of the system and a high-temperature alarm that had been disconnected. (Kosal 2005; Kosal 2006) Whether an operator error or management error we should expect such errors; they are unavoidable. What was avoidable was the wretched condition of the plant that allowed the error to propagate. Supposedly, this was a wake-up call to the industry in the United States. But one doubts that the alarm rang loudly enough.

If there ever was a plant waiting for an accident to happen it was Bhopal in December 1984. Several of the major safety devices were inoperative and had been so for some months, waiting for repairs. There was inadequate instrumentation to detect risky conditions, as had been pointed out in the past by Union Carbide's own inspection teams, and numerous gauges were not calibrated or were inoperative. To save money, the refrigerant had been removed from the key tank that exploded from overheating. The vent gas line, carrying methyl isocyanate (MIC) to an emergency scrubber, leaked most of the MIC to directly to the atmosphere instead. The gas that did get to the scrubber was not neutralized because of a lack of alkali in the scrubber. The vent gas scrubber could only handle five to eight tons of product, while the MIC tank's capacity was seventy tons. Neither the temperature indicator on the MIC tank nor the flare tower for burning off the released MIC was functioning, and the water curtain (high-pressure water sprayers) for neutralizing MIC could reach a height of only ten meters whereas MIC leaked from the vent gas line to about thirty-three meters. The plant staff had been cut back because of business conditions and they were inadequately trained. Management was short of staff and itself lacked experience and training. The alarm system for warning the population was almost nonexistent, and there were no devices to detect the presence or direction of toxic clouds. (Everest 1985; Shrivastava 1987; Ashford et al. 1993, II-9)

Shortly after the Bhopal accident, Union Carbide performed an inspection of a somewhat similar Union Carbide plant in Institute, West Virginia. It found the conditions there to be exemplary and said that such an accident would be impossible. The Occupational

Safety and Health Administration then conducted a thorough investigation and "saw" what they expected—a safe plant. However, Union Carbide's own inspection had disclosed a few things that needed remedy, so it spent about $5 million, principally on a gas leak detection system that would check the prevailing winds and predict where the leak might travel and an increase in the height of a venting tower.

Just eight months after Bhopal, on August 11, 1985, a fairly similar accident occurred at the Institute plant. Aldicarb oxime, dangerous but not as dangerous as the methyl isocyanate at Bhopal, was transferred to a standby tank that was being pressed into service because of some other problems. Unfortunately the operators did not know that this tank had a heating blanket and that it was set to come on as soon as it received some product. Furthermore, a high-temperature alarm on the tank was out of service and a level indicator in the tank was broken, giving no warning that there was excessive temperature and excessive product. With warning systems failing to activate, and the product being mistakenly heated, the tank blew. The new venting tower was not high enough to perform adequately, and the gas warning system was not programmed to react to this particular gas, aldicarb oxime. A few other failures took place, not the least of which was that the number of gas masks in the only protected room, the control room, was set for the normal number of control personnel, but not for the many that flocked there after the explosion to seek protection. The operators survived by lying on the floor and passing the few gas masks back and forth among them in order to breathe, like hippies sharing a "joint." Fortunately the weather and wind was favorable for a gas leak and not many people in the four neighboring communities that were affected were harmed, though about 135 were hospitalized. (Everest 1985; Ashford et al. 1993, II-10)

The Occupational Safety and Health Administration (OSHA) conducted a careful examination of the plant and, as we now can predict, found that this was an "accident waiting to happen." It cited hundreds of "constant, willful violations" in the plant and levied a small fine of $1.4 million, which was later reduced. As

mentioned above, the plant had been inspected by both Union Carbide and OSHA a few months before the accident as a result of the Bhopal "wake-up call" and had been given a clean bill of health. Inspections tell us little. If the plant has been running safely, the managers and agencies such as OSHA will find all is well. If an accident has occurred, the inspectors will find numerous causes for "an accident waiting to happen."

But the Bhopal tragedy lives on. Twenty years later, the tiny $470 million civil suit fine that Union Carbide negotiated with a compliant Indian government still has not been disbursed; indeed, the fund has nearly doubled from accrued interest payments. (Baldauf 2004) Worse still, the Indian government's inaction on ground water contamination is affecting thousands more by leaking poisons, such as heavy metals, nickel, chromium, mercury, and lead, along with other toxic materials such as dichlorobenzene, all of which were used at the Union Carbide plant. The contamination has spread two to three miles from the plant, into the nearby city, where wells are the source of drinking water.

The problem in Bhopal was not a third-world plant failing to measure up to first-world standards; the Institute, West Virginia, case contradicts that. The problem is the institutional structure that regulates the plant and oversees the disbursement of the fines. As flawed as the structure is in the United States, it continues to fail massively in India, which cannot seem to disburse funds to the victims or clean up the mess in Bhopal.

The Bhopal accident is extreme in all respects, but it could happen here. All the elements of that tragedy are replayed yearly on a less grand scale in the United States—the failure of a company to manage its plant safely, its evasion of responsibility, the slight wrist slap it receives from the government, and walking away from the contamination problem. A 2002 Rand report said "[M]any of the chemicals used or produced in plants throughout the U.S. have the potential to match or exceed the 1984 disaster in Bhopal, India. This risk is compounded by the frequent movement of these chemicals, typically by rail, through densely populated areas such as Baltimore and Washington." (Karasik 2002) The threat of a chem-

ical plant failure from a terrorist attack is second only to the threat of a biological weapons attack by terrorists, according to experts. Evidence of al Qaeda's interest in chemical attacks is well-known— copies of U.S. chemical trade publications were found in an Osama bin Laden hideout in November 2001. (Grimaldi and Gugliotta 2001) As always, the source of the tragedy could be nature, industry, or terrorists.

## HURRICANES AND CHEMICAL PLANTS

Before turning to the industrial accident potential and the terrorist potential, we should briefly examine nature's threat—primarily hurricanes in the southeastern United States, where so many plants reside. The issue is explored in a paper by Marc Levitan. (Levitan 2000) Chemical plants abound in areas vulnerable to hurricanes, but direct hits are not common. Hurricane Andrew came close, but it turned aside from the concentration of plants on the lower Mississippi River and was a rapidly weakening category three storm at landfall, with a small eye and small radius of hurricane-force winds. It curved toward Baton Rouge, which has many chemical plants, but its winds of only seventy miles per hour caused only minor damage. However, the category four Hurricane Hugo in 1989 hit the island of St. Croix and produced significant oil spills from damaged petroleum storage tanks and fuel oil tanks; Houston-area plants incurred damage from Hurricane Alicia in 1983; and Katrina, as noted earlier, produced 575 hazardous petroleum and chemical spills. Chemical plants are not safe from hurricane-force winds.

Surprisingly, there are no uniform standards they must conform to. Most of the structures in chemical and petrochemical plants are not addressed in the standards set by the American Society of Chemical Engineers (ASCE). Worse yet, following decades of low hurricane activity, the number of hurricanes increased in the 1990s; the wind speeds were increasing; and the natural barrier of land that weakened the storms was inexorably receding due to development, so that the chemical plants were in effect marching closer

to the Gulf of Mexico each year. Hurricane activity is fairly cyclic, with twenty- to thirty-year stretches of increased activity followed by similar lengths of decreased activity. The recent period of low activity has "led to what appears to be an unfounded expectation that significant damage due to high winds is very unlikely." That, the author of the article says, is a great mistake. As we saw in chapter 2, two large oil companies failed to protect their storage tanks by filling them.

## THE INDUSTRY: BIG AND DANGEROUS

It is not the gross size of the industry that is of concern; we need a lot of chemistry for a lot of better living. For our concerns, it is the inevitably capital-intensive nature of the industry that is important; its capital investment is more than twice the size of all manufacturing. This means huge facilities. Economies of scale abound in this business. A doubling of plant capacity increases the capital costs by only about 60 percent, so bigger is cheaper. (Ashford et al. 1993, III-9) Bigger facilities also mean bigger targets and bigger catastrophes. Accidents have been increasing. Ashford and others wrote in 1993: "A survey of the largest property losses over the past 30 years indicates an increase of 500 percent in the average size of the loss, holding prices constant. . . . A study by the Organization for Economic Cooperation and Development (OECD) notes that, in the Post World War II period, the incidence of major industrial accidents was only one every five years or so, until 1980. Since 1980, the incidence has risen to two major accidents per year." (III-9) The industry claims that it is safer than manufacturing industries, but the workers most at risk are contract workers, hired for short-term jobs from companies that the government does not classify as being in the chemical industry but in various construction industry categories. Since 30 percent of the chemical industry's workforce are contract workers, and they are the most likely to have accidents, the industry's claim is without merit. (Perrow 1999, 361–62)

## GOVERNMENT ACTION

The Chemical Safety and Hazard Investigation Board (CSB) was authorized by the Clean Air Act Amendments of 1990 to investigate chemical accidents and recommend steps to prevent them. However, it did not become operational until 1998. It has no authority to do anything but investigate, but at least that mandate is clear. The legislation that established it no doubt reassured Congress that the investigations would be carried out: "In no event shall the Board forego an investigation where an accidental release causes a fatality or serious injury among the general public, or had the potential to cause substantial property damage or a number of deaths or injuries among the general public." But Congress and the Office of Management and Budget gave it only a $8 million budget in 2004—so small that it could only conduct six to eight full investigations, in a nation where 250 or more persons are killed yearly and hundreds more receive serious injury. (Rosenthal 2004) Nevertheless, its accident investigations provide evidence of the routine failures of plants. I will summarize just one small one here, quoting freely without attribution from their report, to give the flavor of a chain of failures at the plant, the firm that owns the plant, the local authorities, and the chemical industry. (Board 2002)

The DPC Enterprises facility, in Festus, Missouri, had twelve full-time employees. The facility repackages chlorine, transferring it from railroad tank cars into smaller containers for commercial, light-industrial, and municipal customers. On the morning of August, 14, 2002, six workers put the operation on standby and took a break. Twenty minutes later, three men who were smoking outside the building heard a loud popping sound and saw chlorine escaping from a tank car. They rushed to evacuate the area. Three men inside a break room heard a chlorine leak detection alarm and saw chlorine entering the building through an open door and quickly rushed outside. The operations manager pushed the emergency chlorine transfer shutoff button on his way out, but it failed to close the valves on the railcar. An automatic emergency shutdown

system, triggered by chlorine sensors, also failed to close the valves. (We must expect such failures of emergency devices.)

The release continued for about three hours. Emergency responders, with protective clothing, were finally able to stop the leak by crossing through a four-foot-deep yellowish-green fog of chlorine, climbing on top of the tank car, and closing several manual shut-off valves. By that time, some 48,000 pounds of the gas (a small amount, given the tank size) had been released to the environment. While the slow leak continued, fire department personnel notified residents of the Blue Fountain mobile home park and another neighboring area to evacuate. Authorities ordered hundreds of residents, office workers, occupants of an assisted-living facility and a learning center, and students in a local school to remain "sheltered in place" for four hours, staying inside for protection from the toxic gas. Police halted traffic on nearby Interstate 55 for nearly one and one-half hours to keep vehicles from driving into the dangerous cloud. Although light westerly winds kept most of the chlorine away from residential areas, some of the gas likely drifted into the Blue Fountain mobile home park, the CSB said. The accident caused sixty-three people from the surrounding community to seek medical evaluations for respiratory distress; three were admitted to the hospital for overnight observation. In addition, three workers received minor skin exposure to chlorine during cleanup operations after the event. The chlorine also caused trees and other vegetation around the facility and in the plume path to turn brown until the following spring.

How could such a thing happen? The factory manufacturing the hoses "mislabeled" them, so DPC got ones with only a stainless steel braiding, rather one with a more expensive alloy that would resist chlorine corrosion. Both types of hoses look identical, though a simple non-destructive test would have told DPC that it was not getting what it ordered. Neither the fabricator nor DPC had the equipment to do the testing, known as "positive materials identification." It is necessary where different materials look alike and where a mix-up can lead to a highly hazardous event. The mislabeled hose lasted fifty-nine days before bursting.

But there are safety devices that should prevent a runaway accident. The DPC Festus facility had chlorine monitors, safety alarms, and automatic shutoff valves in place that were supposed to stop the flow of chlorine in case of accidental discharge. But the CSB found that four of the five safety shutoff valves failed to close completely when the emergency occurred, due to corrosion and lack of maintenance. Although workers believed they were effectively testing the emergency shutoff system by activating it once each day, they were not required to verify that the valves were fully closed when the system was triggered, a rather obvious requirement.

As DPC employees rushed to evacuate, they notified emergency authorities about the release; but neither the company nor local agencies had an effective system for notifying neighbors. There were no community sirens and no system for making telephone alert calls to area homes, called "reverse 911" calls. Instead firefighters, who arrived about ten minutes after the leak began, had to go door-to-door through neighborhoods with bullhorns ordering residents to evacuate. The company's own emergency training and drills were also inadequate, the CSB said. In addition, chlorine protective gear was stored in the chlorine-packaging building, too close to the tank car unloading station. Once the chlorine leak began, the gear became engulfed and inaccessible, leaving workers no alternative but to flee the plant. The CSB found that emergency responders from Jefferson County, Missouri, were also unprepared for the accident. It took about an hour and a half to assemble all of the county's volunteer hazardous materials team members. The hazmat team determined that chlorine concentrations at the scene were greater than 1,000 parts per million—life-threatening to personnel without proper respiratory equipment. It then took another forty-five minutes to plan entry to the site and don protective suits. By the time team members reached the railcar and were able to shut off the leak, a massive amount of chlorine had been released. (This was a slow leak; imagine if the whole tank burst from a small bomb or a derailment. Or imagine this leak occurring in a large facility, which had gotten the faulty shipment, perhaps creating collateral damage when workers were forced to abandon their stations

for an hour or more. We can better "tolerate" accidents in small plants because there is less collateral damage.)

This was a tawdry series of completely ordinary, prosaic organizational failures. It could happen in a crowded metropolitan area under weather conditions that would spread the toxic gas much more widely, perhaps delaying the response of the hazmat team well beyond the two and a quarter hours it took to notify them and get them there and suited up. This was a small facility, so the damage was limited. But similar errors occur in large facilities with more destructive capability, as documented in the chapter on chemical plants in *Normal Accidents*. (Perrow 1999) What can be done? The CBS issued pious safety recommendations for the facility (it is all they can do), for its owners (the DX Distribution Group), the hose fabricator, the Jefferson County Emergency Management Agency, and the Chlorine Institute, a trade association ("develop an industry-wide system to allow positive identification of chlorine transfer hoses."). (Board 2002) We may be sure it will happen again, and indeed, the DPC had another accident at its repackaging facility, this time in Glendale, Arizona, on November 17, 2003. The absorbent chemicals in a scrubber were exhausted, and fourteen people required treatment for chlorine exposure. (Board 2004)

We are told nothing about any fines in these cases, but we may be assured that they were trivial. Occupational Safety and Health Administration (OSHA) violations in the chemical industry averaged less than $750 in fines during the period from 1972 to 1979, and the average fell during the Reagan administration. The median OSHA fine following a death or serious injury accident in 1990 was reduced, when adjusted for inflation, to less than half of the 1970s median. Furthermore, although relatively large fines have been imposed recently in a few cases—such as the $5.6 million penalties OSHA has proposed for Phillips Petroleum in connection with its 1989 explosion in Pasadena, Texas—even then, the fines are only a minute fraction (typically well less than 1 percent) of the total damage caused by the chemical accident. (Ashford and Stone 1991) Perhaps fines should be scaled to the size of the facility such that the greater the risk the facility presents, the greater the fine for

each injury or violation. This would be an incentive for reducing the size of the targets.

On the night of July 5, 1990, an explosion at the Atlantic Richfield Chemical Company's (ARCO) Channelview production facility generated a fireball and a blast that was heard ten miles away. There were seventeen deaths and five injuries to plant workers. ARCO agreed to pay a $3.48 million fine relating to "willful" violations of federal safety law. A year later, Phillips Petroleum Company was fined $4 million for similar violations that contributed to a 1989 accident at another Houston-area chemical production facility in which twenty-three workers were killed and more than one hundred injured. (Swoboda 1991; Staff 1991b)

## WORST CASES

As ineffectual as it may seem to be for the CSB to do no more than investigate and tell everyone to play safe, merely making its six to eight careful investigations a year will do some good. Much more good was done when, in 1999, the CSB conducted a census of chemical industry accidents, the first ever. It covered the period from 1987 to 1996. The CSB was uncertain that all the regulations imposed on the industry were having any positive impact on health and safety. Chemical accidents happen more frequently than most of us would ever imagine, they observed, and occur in every state—and not just in chemical plants but in other kinds of plants, including those that provide our water supply—and on railways, highways, and waterways. In ten years, there were 605,000 reported chemical incidents (many additional accidents are never reported), and almost one-third caused fatalities, injuries, evacuations, or property damage. There were more than 60,000 incidents per year and almost 2,600 deaths in the ten-year period. Unfortunately, the report was never published. It exists as "The 600k Report: Commercial Chemical Incidents in the United States, 1987–1996, Special Congressional Summary," and I have drawn on Lee Clarke's summary for the data given above. (Clarke 2005, chap. 4)

But another report did make an appearance for a time on the World Wide Web that is more relevant to our concerns than just the number of accidents. It asked: What are the hazard potentials of chemicals? As a result of the 1990 Clean Air Act, facilities with significant amounts of chemicals had to prepare worst-case scenarios, including projections of potential consequences, and report them to the Environmental Protection Agency (EPA), which consolidated them and issued the report. In 1999, Congress agreed with chemical companies to restrict the Internet availability of the EPA's worst-case scenarios for individual plants on security grounds, but the scenarios remained on the Web for some time, and a remarkable *Washington Post* story summarizes some of the findings. (Grimaldi and Gugliotta 2001)

A single railcar of chlorine, if vaporized near Los Angeles, could poison four million people. Four million people could be harmed by the release of 400,000 pounds of hydrogen fluoride that a refinery near Philadelphia keeps on hand. A chlorine railcar release near Detroit would put three million at risk. Union Carbide's Institute, West Virginia, plant near Charleston once again has 200,000 pounds of methyl isocyanate—the chemical that did so much damage in Bhopal—which would threaten 60,000 people. (Union Carbide had removed the methyl isocyanate storage from the Institute plant after the Bhopal accident, but the Institute plant was reported to be storing it again in 2001.) And close by New York City, a plant in New Jersey has 180,000 pounds of chlorine and sulfur dioxide, which could create a toxic cloud that would put twelve million people at risk. (Grimaldi and Gugliotta 2001) A terrorist attack on such facilities would not require the sophistication needed for an attack using a biochemical weapon.

The chemical industry spokespersons dispute the findings, noting that for the damage figures to be accurate, the plume of, say, chlorine gas, would have to spread evenly over the area, whereas it would possibly be a cone that covers only a part of the area. The EPA and the DHS have agreed with this reservation, but the Government Accountability Office, in addressing the issue, points out that a "worst case" was assumed to involve the release from only

one source: the most deadly one, such as the biggest tank or the most heavily inventoried process. But the GAO points out that other parts of the site might be disabled by the failure of the one source, bringing much more inventory into play. An attack that breached multiple chemical vessels simultaneously could result in a larger release with potentially more severe consequences than those outlined in worst-case scenarios. (Stephenson 2005; GAO 2004)

It is worth visiting the Web page of D.C. Councilmember Kathy Patterson on hazmat transport regulation http://www.dccouncil .washington.dc.us/patterson/pages/prinfo/HazmatPhotos.htm. It shows the calculated plume from a chlorine release in the Washington DC area, which the U.S. Naval Research Lab estimates could kill 100,000 people in half an hour, and it is enormous. It also shows photos of chlorine tank railroad cars parked near the Capitol, traveling next to highways in the Capital area, with graffiti on them (indicating their accessibility); a Chlorine Institute estimate of 180,000 pounds of release in ten minutes traveling 14.8 miles downwind; liquefied propane tank cars on railroad bridge at Seventh Street, SW, in Washington DC; and a parked chemical tank car at Ninth Street, SW. It also quotes a railroad engineer saying a handheld grenade launcher, rifle, or bazooka could easily puncture a railcar of deadly gasses, killing thousands within minutes. And, of course, a commonplace railroad accident could do the same.

One can easily see how an accident, storm, or terrorist attack would release many hazmats from their containers through destructive explosions or fires. Or, take the vinyl chloride accident on April 23, 2004, where five workers were killed, four towns evacuated, several highways closed, a no-fly zone declared, and three hundred firefighters from twenty-seven surrounding communities battled the flames for three days. It took that long because the company, Formosa Plastics, ran the wells that provided the town's water and the shock wave of the blast (a hundred-foot fireball) disabled the town's water supply. Thus, the figures released by the EPA are likely to be underestimates, as frightening as they already are. (Steingraber 2005)

One last disaster, this one in Toulouse, France, occurred two

weeks after 9/11. A massive explosion at a fertilizer plant killed thirty people and injured another 3,500. At least ten of the dead were people who lived near the plant; and 11,000 homes were destroyed and an additional 16,000 structures were damaged. The explosion registered 3.4 on the Richter scale. It is unwise to live close to a fertilizer plant. But the most interesting aspect of the accident is that the French judge who presided over the inquiry charged the company with deliberately encouraging investigators to find terrorists responsible in an effort to divert attention from negligent safety practices and insufficient storage facilities at the plant. The editor of the leading industry publication in the United States sagely noted "it will enormously help the chemical industry's reputation if it becomes possible to prove that the blast was not an accident." (Hunter 2001)

## THE INDUSTRY RESPONSE

Of course the chemical industry is aware of these dangers. Accidents are very expensive and might provoke costly governmental regulations. The industry has taken some steps over the years to mitigate the risk. The principal organization representing chemical manufacturers was founded in 1972 and initially called the Chemical Manufacturers Association. It changed its name to the more friendly one of American Chemistry Council (ACC) in 2000, and has played an active role in promoting safety among the members, claiming that its members have spent $2 billion on safety in recent years. It represents approximately 140 companies that manufacture basic chemicals in some 2,000 facilities in the United States, accounting for more than 85 percent of basic chemical production in the country. In 1988 it borrowed a safety program that Canadian firms had developed, expanded it, and trademarked it as the Responsible Care Management program. The program has been praised by the U.S. Environmental Protection Agency. All its members have voluntarily agreed to abide by the plan, and citing the competitive advantage afforded to those that do not, the ACC

would like to see legislation to require the remaining 15 percent of chemical manufacturing companies to abide by the Responsible Care Management program. (Jenkins 2005, 2, n. 3)

But critics of the program note some important flaws. According to the ACC document regarding verification of inspections performed on plants, the companies can select the verifiers, for example, firefighters and police in the local community, security consultants they hire, and local insurance auditors. This could mean that the company picks the "low-hanging fruit," verifiers that may not be qualified and are most easily influenced by the local plant, which is generally the most powerful economic entity in the community. The GAO, in its report "Protection of Chemical and Water Infrastructure" (Jenkins 2005) raises this issue, as do critics from public interest organizations. The latter groups want a bill that has the federal government either do the verification or provide a list of approved verifiers according to national standards for verification. The president of the ACC disagrees, and was quoted shortly after the 9/11 attack as saying, "Additional regulations, stronger enforcement—that isn't going to do the trick," he said. "What you need is the industry stepping up on its own, preventing the worst from happening." (Grimaldi and Gugliotta 2001)

"Stepping up on its own" did not impress the GAO. Not only was it concerned about the independence of the verifiers, worse still, the Responsible Care Management system certification "does not require third parties to verify that a vulnerability assessment was conducted appropriately or that actions taken by a facility adequately address security risks." (Jenkins 2005, 18, 27 table 2, n. b) It is like taking a test and then grading it yourself. Companies volunteer to join the ACC and abide by its program, but can do an inappropriate assessment and make no changes. This would make the ACC claim that all their members follow the guidelines rather meaningless. Some evidence of this appears in a recent survey by the Paper, Allied-Industrial, Chemical and Energy Workers International Union (PACE). It found that, according to a journalist with *Chemical Engineering News,* fewer than 17 percent of the industrial chemical facilities have enacted "fundamental changes that

would lower the impact of an accident or attack by making chemical processes inherently safer or by storing smaller amounts of hazardous materials on-site." The article also reported: "What actions were taken were mainly in terms of guards and security (76%), and only 17% dealt with making processes inherently safer or reducing hazardous material storage." (Johnson 2004)

The GAO asked ACC members for volunteers to be interviewed by the GAO staff for a study. Ten plants agreed. The results were mixed, even for this self-selected sample (presumably the most progressive): in only seven of the ten were they making process or inventory changes that would reduce hazardous chemicals on site, and only five of the ten had established system redundancies such as backup pumps, backup power systems, and storage capacity. (Jenkins 2005, 21) That seven of the ten were reducing storage is good news, but presumably these are the showcase plants; the PACE survey found nothing of the sort.

Some of the company's problems deserve sympathy, and tell us something about interorganizational complexities. Three of the ten had trouble trying to install fences because the Army Corps of Engineers controlled the adjacent property. Part of the corps' legal mandate is to protect the wetlands, so building a fence required a permit and the corps was not cooperating. (29) In two cases the chemical company complained that it had no control over the security practices of the railroads that served the facility, whose security levels were below those the plants; two other chemical plant officials complained that the contractors they hired for periodic major maintenance jobs did not have an appropriate level of background screening. (22) (They could hire and train their own workers, of course, but contract workers are much cheaper and nonunion.)

Nevertheless, the efforts of the ACC do not match those of the U.S. Coast Guard, which is responsible for chemical facilities in ports and harbors. The Coast Guard inspected 2,900 regulated facilities in the last six months of 2004, took 312 enforcement actions against owners or operators, and imposed "operational controls" over twenty-nine facilities, including suspending their operations. (Jenkins 2005, 8) Their actions included unscheduled spot checks

as well as scheduled inspections, developing detailed checklists for the inspectors to use, reviewing the enforcement actions taken by inspectors, and training them. (30) Nothing comparable occurs in the voluntary program of the rest of the chemical industry. If the Coast Guard can do it in six months for 2,900 facilities, it should not be hard to fund another government agency such as the EPA or OSHA to do an equally careful job on 4,000 dangerous chemical plants and storage sites.

The effectiveness of the chemical industry's voluntary Responsible Care Management program has been questioned by those examining the data on emissions by members of the program versus those organizations that have not joined it. The results are discouraging, and the analyses suggest a "gaming" of the system. Participation in the program means that the EPA is not likely to check on the emissions from the plants. In a survey of 3,606 facilities involving 1,500 firms over the period from 1987 to 1996, the researchers found that while emissions overall declined significantly, the firms that were not members of the program reduced their emissions considerably more than firms that were members. The report says: "Our data provide no evidence that Responsible Care has positively influenced the rate of improvement among its members. Indeed, we found evidence that members of Responsible Care are improving their relative environmental performance more slowly than nonmembers." (King and Lenox 2000, 709) Belonging to the much acclaimed and EPA-praised voluntary program seems to have encouraged the members to make fewer improvements then they would have if they knew the EPA was watching. Similar findings appear with studies of other industries who have been allowed to engage in voluntary self-regulation, a practice that started in the 1990s and has steadily increased. (Delmas and Keller forthcoming; Delmas 2000; Harrison 1999; Welch, Mazur, and Bretschneider 2000) What goes for emissions is very likely to be the case with self-regulation to improve security.

In response to the 9/11 attack, Senator Jon Corzine (D-NJ) introduced a bill in 2002 requiring the federal government to set up security standards for chemical plants and to have plants not just

enhance their security but reduce the amount of hazardous materials in storage and make efforts to substitute safer materials. The ACC strongly opposed it, favoring voluntary actions by the chemical manufacturing community, and the bill died in committee. Corzine attributes the defeat of the bill to a strong lobbying effort. (This may account for the unfortunate fact that since taking office as governor of New Jersey in 2006, he has made no move impose such standards on the facilities in his own state, even though the governor has to power to do so.) So did the EPA administrator of the time, Christine Todd Whitman. In her book, *It's My Party Too* (2005), she charged that industry lobbyists worked with key Republican lawmakers to sabotage new security regulations for chemical plants after the 9/11 attacks. One newspaper account of her battles with the White House indicates that she and Tom Ridge, when he was heading up domestic security from the White House, before DHS was born, worked on a modest effort to require high-risk plants—especially the 123 factories where a toxic release could endanger at least one million people—to enhance security. But industry groups intervened, warning President Bush's political adviser, Karl Rove, that giving new regulatory power to the Environmental Protection Agency would be a disaster. As noted earlier, Rove wrote in a reassuring letter to the president of BP Amoco Chemical Co.: "We have a similar set of concerns." In her book, Whitman confirms the White House veto of the proposal but singles out two people from her own party for blame, Senator James Inhofe (R-OK) and Representative Billy Tauzin (R-LA) "Although both Tom and I agreed such legislation was necessary, strong congressional opposition, led by some Republicans on the Senate Environment and Public Works Committee and the House Energy and Commerce Committee, to giving EPA even this modest additional statutory authority made it difficult to secure administration support." (Lane 2005) But the pressure has been increasing with strong editorials in the *New York Times* and other papers and now the ACC calls for federal legislation. (Staff 2005h)

A weak bill was still under consideration in March 2006, advocating uniform national standards. But compliance would not be checked by a government agency, or necessarily by a nonprofit pro-

fessional association, but could be validated by a for-profit organizations. Still, DHS secretary Michael Chertoff seemed opposed to
allowing some states such as New Jersey to set higher standards
than the national ones, and he was opposed to forcing companies
to switch to less dangerous chemicals even where this was feasible.
This annoyed a spokesperson for the Greenpeace Toxics Campaign
who pointed out that there were forty-five sewage treatment plants
in urban areas, as well as power plants and refineries in a number
of states, that used extremely dangerous chemicals like chlorine
gas and hydrochloric acid while there were widely used alterna-
tives that did not impose excessive costs. (Lipton 2006c)

## THE MEDIA COMES ABOARD

The next major actor in the institutional framework of firms, public interest groups, Congress, and federal agencies such as the
Chemical Safety Board, EPA, and GAO is the media. An enterpris-
ing reporter, Carl Prine, of the *Pittsburgh Tribune-Review* began
probing security at chemical plants six months after 9/11. Chemical companies had been warned by the government that they were
potential targets. Prine visited sixty plants all over the country and
found that despite the government warning, security was extremely
lax, even nonexistent. He walked onto plant grounds and up to
tanks of dangerous chemicals and took pictures of them. Rarely
was he stopped or questioned; more often, if anyone noticed him,
it was to give him a cheerful wave. (Prine 2002) Of course, he was
not dark-skinned and wearing a turban. People like that were being
arrested in Washington DC and other parts of the country for taking photographs of river views and monuments, as a result of the
Patriot Act. His investigation was fairly widely reported, but Prine
doubted that much had changed after his stories ran, and about two
years later he again went into the field. He teamed up with CBS's *60
Minutes* crew and investigated fifteen plants around Pittsburgh and
Baltimore (CBS went on to four more states). (Prine 2003)

Prine's findings are very depressing, but given this book's view
that for-profit organizations are weak vessels when the public good

is concerned and that even when they try hard the failures will be many, we should not be surprised. Some plants had installed the usual fences, cameras, lighting, and additional security guards, but to no avail. Prine and his photographer went through unlocked gates, holes in the fencing, or down railroad tracks, past inattentive guards, and up to huge tanks and railroad cars filled with extremely dangerous hazmats, occasionally joking with workers.

The facilities included the warehouse of the grocery giant Giant Eagle, where the journalists went through a fence hole and up to a tank holding 20,000 pounds of anhydrous ammonia, a coolant for refrigeration, which could put nearly 43,000 people—including children in twenty-four schools—at risk of death, burns, or blindness, according to company filings with local emergency planners. At the mammoth Sony Technology Center in Westmoreland County, Pennsylvania, an unsecured gate, distracted guards, and unconcerned employees let them reach 200,000 pounds of chlorine gas. (Recall the damage done by the leak—not a full breach of the tank—of just 48,000 pounds that the Chemical Safety Board investigated.) No one stopped them as they touched train derailing levers, waved to security cameras, and photographed chlorine tankers and a nitric acid vat. If ruptured, one Sony railcar could spew gas thirteen miles, endangering 190,000 people. A water treatment plant spent more than $100,000 on electric gates, cameras, and identification badges, but the reporter went through a fence hole and an unlocked door that led to *twenty tons* of chlorine gas. The director of the plant was very concerned with the security breach; he noted that terrorists can hit natural gas or electricity supplies and people will survive, but that they can't live without water. (True, but it is hard to live *with* twenty tons of released chlorine gas.) An industrial chemical distributor erected fences, instituted round-the-clock guards, installed cameras and even fortified its river dock. But federal safety laws would not allow the railroad to fence off the track where a chlorine tanker parks daily. The reporter and the CBS crew were able to make four undetected trips up the rails to *ninety tons* of chlorine gas. Railroad cars filled with toxic substances are so easy to get to that they are frequently covered with graffiti. (Kocieniewski 2006)

The journalists then went to the companies which, according to workers, put more effort into making sure toilet paper was not stolen than they did in protecting the public, and most company officials declined to comment. They then went to Tom Ridge, head of the DHS, who was concerned but also said that he was optimistic that long-term federal reforms will provide protection.

Senator Corzine, whose state of New Jersey has more than its share of chemical plants, joined the CBS camera outside a chemical facility that was a couple of hundred feet below a well-traveled highway overpass, near the New York border and Manhattan. The footage shows them walking up to an unguarded tank, its gate ajar. According to government records, nearly *one thousand tons* of deadly chlorine gas are stored here—the first agent ever used in chemical warfare during World War I. Chlorine gas does not burn or explode itself, but a full release following, say, an explosive charge on just one of the tanks might create conditions that could release the gas in others (e.g., abandonment of monitoring controls, enforced operating errors); stranger things have happened in this industry.

"This is one of the main thoroughfares for commuters who come in and out of New York City every day," says the senator on the tape. "You know, it looks to me like you could drive a truck through some of these fences if you wanted to pretty quickly." According to the plant's worst-case estimate, the number of people at risk—those living within a seven-mile radius of the plant—was fourteen million. CBS then interviewed the head of the American Chemistry Council, telling him of their visit; curiously the official expressed reassurance. He said what they found was "totally unacceptable," but that "it underscores the point that you and I have been discussing—that we work every day on being better at security. One security breach in one facility or several facilities is unacceptable to us." It was an exception, he insisted, but the reporters had found dozens of such exceptions. (CBS News 2004)

The initial *Pittsburgh Tribune-Review* story appeared in 2002, the next Prine story and the CBS *60 Minutes* report in 2004. In the spring of 2005, the tale continued. A *New York Times* reporter and a cameraman looked over the two miles of New Jersey that are said

to be the most dangerous two miles in America, home to three major oil and natural gas pipelines, heavily traveled rail lines and more than a dozen chemical plants, a prime example of the concentration we are concerned with in this book. A congressional study in 2000 by a former Coast Guard commander deemed it the nation's most enticing environment for terrorists, providing a convenient way to cripple the economy by disrupting major portions of the country's rail lines, oil storage tanks and refineries, pipelines, air traffic, communications networks, and highway system. The reporters focused on a particularly deadly plant, a chemical plant that processes chlorine gas. It remained loosely guarded and accessible. Dozens of trucks and cars drove by on the highway within one hundred feet of the tanks. The reporter and photographer drove back and forth for five minutes, stopping to snap photos with a camera the size of a large sidearm, then left without being approached by plant employees. (Kocieniewski 2005)

In 2005, the federal Department of Homeland Security cut New Jersey's financing to about $60 million from $99 million in the previous year. (Kocieniewski 2005) (Because of congressional politics, Montana, with almost no terrorist targets, gets as much federal aid for security as New Jersey, one of the top three most vulnerable states.) One would think that Michael Chertoff, the new head of the DHS would be sympathetic to the state's situation because he is a native of Elizabeth. But when he visited New Jersey during a terror drill at this time (April 2005), he was noncommittal about restoring cuts. He told the reporter, "Frankly, it's not a matter of spending a great lot of money, it's a matter of taking resources we have and having a plan in place so we use them effectively."

New Jersey officials see it differently. They say that the cuts will force them to reduce surveillance of possible targets, cancel training sessions for first responders and counterterrorism experts, and forestall the purchase of equipment to detect chemical, nuclear, or biological agents. They would have to scale back plans to fortify storage facilities and rail lines near the Pulaski Skyway, an area known as Chemical Alley. A DHS spokesperson was not worried; the DHS had visited more than half of the nation's three hundred most dangerous plants and had "urged" the companies to enhance

perimeter security and to switch to less hazardous chemicals and processes. In a remarkable non sequitor, the newspaper quotes her as saying that as a result of this effort, she believes North Jersey is "one of the safer areas because it has received the most attention in terms of protective measures." (Kocieniewski 2005) Perhaps that is why federal funding was cut. Observers have noted that New Jersey is a "blue" state.

The legislative battle is illuminating. Senator Corzine introduced his Senate bill in 2002, and Senator Inhofe introduced a competing bill in May 2003, which had wide support from the chemical sector and the security industry. The Corzine bill emphasized using "inherently safer technologies." The Inhofe bill did not; it focused strictly on physical security. As noted, the Corzine bill required companies to submit response plans to the government, which would review them; the other bill only said that the government could request them at times and for places that the government officials deemed appropriate. The first bill designated the EPA as the lead agency because of its experience with existing accident-prevention requirements and would do the implementing in coordination with the DHS. But the Inhofe bill limits any other agencies than the DHS from involvement; they can only providing "technical and analytic support" on request by DHS and specifically bars them from any "field work." The Corzine bill would set standards "to eliminate or significantly lessen the potential consequences of an unauthorized release"; the other would set no standards, such that extending a guard's hours would suffice to "reduce vulnerability."(NRDC 2003) In July 2002 the Corzine bill was adopted unanimously by the Senate Environment and Public Works Committee (EPW). But more than thirty trade associations, led by the ACC, signed a letter opposing it, and within a month seven members of the EPW reversed their position, and the bill died. (Jarocki and Calvert 2004) The Inhofe bill also died in committee the next year. The ACC supported the Inhofe bill, and President Bush later excluded the EPA from the issue of plant security.

But all may not be lost. Encouraging news appeared in June 2005. Back in 2003, the administration favored legislation restricted to voluntary standards self-assessed by plant owners. But in 2005, in

testimony to congress, Robert Stephan, a top deputy to DHS head Chertoff, said, "The existing patchwork of authorities does not permit us to regulate the industry effectively. . . . It has become clear that the entirely voluntary efforts of those companies alone will not sufficiently address security for the entire sector." Federal standards may finally be forthcoming, but opposition was voiced at the June hearings. A representative of the American Petroleum Institute said, "Industry does not need to be prodded by government mandates. . . . Chemical security legislation would be counterproductive." (Lipton 2005) It also seems likely that any legislation will focus primarily on security and hardly at all on reductions in volume and toxicity.

## WILL INSTITUTIONAL PRESSURES CHANGE THE INDUSTRY?

It will take tough legislation to capture the attention of the chemical industry, which is so well situated with their massive campaign financing that it will be difficult to get Congress to act. However, it is not impossible, since Congress has acted in the past as a result of public pressure and especially that of environmental groups. Andrew Hoffman explores the development of environmental concerns by the petroleum and chemical industries from 1970 to 1993, arguing that they moved from considering environmental concerns as heresy to accepting them as dogma. Over just three decades, the chemical and petroleum industries "moved from a posture of vehement resistance to environmentalism to one of proactive environmental management." (Hoffman 1997, 6) His enthusiasm for the changes in the industry is premature, I believe, since major instances of falsifying or hiding data continue to appear, and the industry even goes to great lengths to discredit critical scholarly works that expose their record, as we shall see. But his account of the pressures on the industry is revealing and deserves a summary. Public pressure can make a difference, so all is not lost.

In the 1960s, there was hardly any serious legislation concerning

emissions, pesticides, or carcinogenetic substances. From 1960 to 1970, government regulation was minimal and environmentalists had little influence, so the chemical and petroleum industries were free to establish their own conception as to what safety and environmentalism meant. The oil industry flatly denied that emissions were harmful and that oil spills damaged the ecosystem, and said that no regulation was needed. Even the publication of Rachel Carson's *Silent Spring* in 1962, probably the most influential environmental tract ever published, had no effect in the 1960s on the practices of chemical firms, which denied that pesticides did any harm.

But oil spills, air pollution, and the dangers of Agent Orange—which was used in the Vietnam War and affected our troops—kept emerging. The "sixties generation" was stirring things up, and protest movements appeared, bringing the first Earth Day celebration in 1970. Local and state regulations were being enacted with confusing and contradictory requirements, so the chemical industry called for a national policy on chemical pollution. The result was two federal acts: the Clean Air Act of 1970, targeting the oil industry, and more significant, an act establishing the EPA, signed by President Nixon in 1970. The EPA was at first welcomed by the chemical industry, to bring "order out of confusion." But the only appreciable impact of the 1960s' growing environmental concerns on the industries was that some firms formed their first environmental departments. These had little power in the organizations and focused strictly on legal requirements that the new federal laws were establishing. (Hoffman 1997, 51–55)

The formation of the EPA was a major restructuring for the federal government at the time, bringing in people from many other organizations that were concerned with air, water, pesticides, radiation, and solid waste. Nearly six thousand employees were moved to the new department, small in comparison with the Department of Homeland Security but large at the time. That it worked much better than similar reorganizations may have a lot to do with its strong-minded first administrator, William Ruckelshaus. During its first sixty days, the agency brought five times as many enforcement actions as the agencies it inherited had brought during any

similar period. Hoffman looks askance at this "penchant for strong enforcement," and says it was justified to establish agency credibility, rather than saying it was needed in its own right. Unfortunately, he says, it established an "adversarial, 'command-and-control' type of relationship between government and industry." (65) Given industry's denial that its pollution was at all harmful and its reluctance to change, there might have been no other alternative.

Industry responded to the outpouring of a string of tough laws by bringing its own lawsuits against the new agency, filing forty to fifty suits a year between 1976 and 1982 and charging that the government was the "biggest predator of them all." (68, 75) President Reagan came to power on a antigovernment platform and in 1981 appointed a new director, Ann Burford Gorsuch, who proceeded to cut the agency's budget and reverse its rulings. Industry was delighted, but the reversal was so extreme that Reagan was forced to fire her in 1983, her assistant resigned, and another key official in the EPA went to prison for lying to Congress. Industry was outraged but found it hard to sustain the outrage when, in 1984, the horrendous Bhopal disaster occurred.

More legislation, and more pollution, such as the *Exxon Valdez* oil spill (1989) followed. Environmental expenditures by corpo rations increased at a steady rate of $250 million per year from 1973 to 1993. Hoffman does note that this was not voluntary on industry's part; it was because industry was "forced to react." (6) By 1992, firms in both industries were forced to spend 10 percent of their capital budgets on environmental projects, but even so, the chemical industry remained far and away the number one polluter and the petroleum industry, number seven, according to the EPA. (10)

I have not dealt with the chemical companies' problems with toxic consumer products, or the exposure of workers to dangerous substances. There are disasters here, of course, but not the catastrophic ones that concern us in this book. However, a number of recent exposes and lawsuits suggests that the rather benign neoinstitutional viewpoint expressed by Hoffman—that the public and the government will eventually compel sound environmental prac-

tices that large companies will not only accept but the companies will become "proactive" environmentalists—leaves something to be desired. One such case is worth reviewing, since twenty chemical companies, led by the ACC, broke new ground in defending themselves from the charge that they knew they were poisoning their employees and perhaps poisoning a vast number of the U.S. population. The aggrieved widow of a chemical worker brought a lawsuit that could have devastating consequences for the companies since, if successful, it would be trigger massive class-action suits. An important part of the evidence came from a book by two academics. The companies charged that the research used by the plaintiff was flawed and the two eminent university professors that conducted it had engaged in unethical behavior, so the evidence should be thrown out.

The evidence came from a warehouse full of company documents that a lone lawyer had collected in the process of filing an earlier suit. Apparently, the companies intended to drown this lawyer and his tiny firm in documents and discourage the suit. They succeeded and the lawyer gave up. But he offered the documents to the two historians, Gerald Markowitz and David Rosner, who published a book, *Deceit and Denial: The Deadly Politics of Industrial Pollution* (2002) based on the files. The book got rave reviews in the scientific press. Another suit was brought by the wife of a worker who died of an extremely rare disease caused by exposure to vinyl chloride monomer on the job. The chemical companies named in the suit charge that the research of Markowitz and Rosner is "not valid" despite being based on company documents that, for example, acknowledged they risked being charged with "an illegal conspiracy" by concealing the vinyl chloride–cancer link. The companies charged that the publisher's review process, even though using an unprecedented eight reviews, was "subverted," and that the two authors "frequently and flagrantly violated" the code of ethics of the American Historical Association. The companies subpoenaed all the records of the publishers (the University of California Press and the Milbank Memorial Fund), and even went so far as to subpoena five of the eight reviewers of the manuscript.

They subjected Markowitz to five and a half days of grueling deposition conducted by fifteen different chemical companies. A key question: Had the reviewers checked all 1,200 footnotes? They had not checked all of them; it is not a requirement for reviewers. But attorneys for PBS and HBO, both of whom ran specials on cancer caused by chemicals in consumer products such as hair spray and vinyl food wrap did check them. (Wiener 2005) The case was settled in 2006, but Markowitz and Rosner were not involved in the settlement. Markowitz is set to testify in another vinyl chloride plant worker suit, but the chemical companies have moved to have his testimony excluded, again on the grounds that his book violated the ethics of the American Historical Association. (Personal communication)

The actions of the ACC and its members might not be quite as bad as denying that pollution and pesticides harm people, as they did forty years ago, but they seem not to have come very far in "embracing environmentalism," as Hoffman claimed.

The advances that have been made, and there are many, have been in limited areas: obvious pollution, and exposure of workers and communities to persistent or occasional hazmats. Hopefully the next phase in Hoffman's sequence of periods will be a concern with large-scale disasters that kill many people or contaminate large areas. The origins of the disasters could be, as we have relentlessly repeated, natural, industrial, or deliberate in source. The first two still remain under the radar, even for most activist groups. But 9/11 has brought the third, terrorism, into view. And as yet, it seems that the institutional field has not mobilized enough to move the chemical industry from "heresy to dogma" regarding the terrorist risks to which they are subjecting us.

## CAN REDUCTIONS AND SUBSTITUTIONS BE MADE?

Are we being realistic in calling for reductions and substitutions? Is the size of plants and of chemical companies the basic problem? There is no doubt that chemical companies are among the biggest in the world, and many of those operating in the United States are

foreign owned and certainly global in scope. The size of the companies operating in this country is already so enormous that increases in recent decades are beside the point. I could find no data indicating any increase in the size of storage facilities, per se, in the last fifty years or so, but again, the amount of hazmats stored at each site is already so large that no such data are required. However, such an increase seems quite likely to have occurred. A rather old study of the size of chemical plants found a startling fivefold increase in size in only twenty-five years, from 1957 to 1982. (Lieberman 1987) It seems likely that the larger the plant, the larger the number of storage vessels, and probably the larger the vessels themselves.

A study by Nicholas Ashford and associates in 1993 provides evidence of the storage/size problem. Analyzing accidents, Ashford found that storage releases were far more severe than releases from processing, valves and pipes, disposal, and other activities. For primary processors, storage releases were six times greater than the next category (valves and pipes); for secondary producers, twenty times larger. (Ashford 1993, III-15) Nevertheless, Ashford does not think that risks will be reduced by dispersing storage tanks because of the risk of transportation accidents. While favoring primary prevention by using less-hazardous materials and safer processes he does not see much future for inventory reductions. "Some reductions in inventory of hazardous substances, while heralded as primary prevention, may simply shift the locus of the risk and increase the probability of transport accidents." (Ashford 1993, i, VII-16)

This is a widespread belief that should be challenged, since it is true only in some cases but not in most. First, reductions of inventory do not necessarily mean more transport of inventory. Inventory reduction can occur through closed-loop production, where the only hazardous materials are in the production loop rather than in storage, and more efficient processing. Second, the location of the risk is quite important. If sixty tons of hazardous materials are stored in six places rather than one, the probabilities of an accident in *each* of the six is much lower than the probabilities of an accident in the one concentrated storage. If there is an accident in one of the six, the damage will be only one-sixth of that of an accident

in the concentrated one, and probably a great deal less because the collateral damage of one big tank accident on adjacent facilities will be a bigger event and probably in a bigger plant. Finally, the issue of transport accidents is not clear-cut. If each of six plants requires ten tons of the hazardous material to be in storage, and it is made on each site, there is no transportation hazard but there is a substantial deconcentration. If they buy it from a producer, the shipping hazards would probably be same whether the producer shipped it to one big plant or six small ones. (It could be even less in the latter case, since smaller containers would be more likely to be used.) Production economies may well be affected by dispersed storage, but these should be compared with the substantial reductions in the risk of natural, industrial, and terrorist dangers.

One might argue that the bigger the firm, or the plant, the more notice it will attract, and thus it will spend more resources on safety than will smaller plants. The limited evidence we have suggests the opposite. Paul Kleindorfer, of the University of Pennsylvania, found that the larger the plant, the greater the number of unwanted releases, even controlling for size of inventory. (Kleindorfer et al. 2003) Professor Don Grant, of the University of Arizona, found the same thing for chemical company plants, and for the size of chemical companies themselves. The bigger the plant, the more the pollution; the bigger the organization that owns the plants, the more the pollution. (Grant and Jones 2003; Grant, Jones, and Bergesen 2002) These studies dealt with accidental releases or just sloppy procedures, but they are related to catastrophic potentials. Whether it is that the size of the facility makes it more unmanageable and more prone to accidents, or that the power that comes with size makes it less concerned with damage to its reputation or community inhabitants, and thus it neglects safety, is not known. But *big is not necessarily safer.* A fire or explosion in one vessel is likely to spread to others in the plant, and a release of a toxic substance is likely to disable employees who are needed to keep other processes stable or, as we saw above, disable the water supply needed to fight the fires. Large sections of chemical plants are typically destroyed by collateral damage. Their very size is a hazard, and the size of the storage the greatest hazard.

We are not likely to either break up big firms or require them to have three or four dispersed plants instead of one concentrated one. But a reduction in storage volume and toxicity is certainly feasible, and have little or no economic impact. Here are a few examples, copied verbatim from the U.S. Public Interest Research Group Education Fund publication, "Irresponsible Care." (Purvis and Bauler 2004, 10)

- A Massachusetts state law requires companies to disclose the chemicals used by their facilities, including the amounts on site, transported in products, released to the environment, and generated as waste. Companies also are required to produce toxics-use reduction plans. As a result, between 1990 and 1999, facilities reduced their use of toxic chemicals by 41%, while at the same time production increased by 52% and companies saved $15 million.

- Reductions in storage volume and toxicity may have played a role in the reduction of releases of carcinogenic chemicals by 41% between 1995 and 2000, as a result of the federal Toxic Release Inventory program. The program requires several industry sectors to report the toxic chemicals they release into our air, water, and onto our land.

- In Washington, DC, the Blue Plains Sewage Treatment Plant switched from volatile chlorine gas, which could have blanketed the nation's capital in a toxic cloud, to sodium hypochlorite bleach, which has almost no potential for an off-site impact. In the wake of September 11th, 2001, the facility completed the switch in a matter of weeks. The expected cost to consumers will be 25 to 50 cents per customer per year.

- In Cheshire, Ohio, American Electric Power selected a urea-based pollution control system rather than one involving large-scale storage of ammonia that would have endangered the surrounding community.

- In Wichita, Kansas, the Wichita Water and Sewer Authority's sewage treatment plant switched from using chlorine gas to ultra violet light in its disinfection processes. The plant expects to save money in the long run as a result of the change,

as there is about a 20% anticipated cost savings in energy costs versus chemical costs.

• In New Jersey, more than 500 water treatment plants have switched away from or are below threshold volumes of chlorine gas as a result of the state's Toxic Catastrophe Prevention Act.

• In 2003 in Wilmington, California, the Valero Refinery switched from hydrofluoric acid, which when released forms a toxic cloud that hovers over surrounding communities, to modified hydrofluoric acid, which is less hazardous. This change was largely due to decades of community pressure after a devastating accident at a near-by refinery in the area.

Hydrofluoric acid was involved in a 1987 release in Texas City that sent more than one thousand people to the hospital and forced three thousand residents to evacuate their homes for three days. (TexPIRG 2005) (That unfortunate city also had a horrendous series of explosions in 1947 that leveled large parts of the city and killed thousands.) Refineries could switch to sulfuric acid, which has a lesser off-site threat, or modified hydrofluoric acid, which is less toxic and which the Wilmington, California, refinery switched to after an accident and community pressure.

Nicholas Ashford and his research team, writing in 1993, report on other instances of substitutions that are encouraging. Monsanto cut its total volume of highly toxic gases in storage facilities throughout the world in half by shifting to just-in-time deliveries of raw materials. Rohm and Haas replaced its batch processing system to continuous processing, cutting the inventory of toxic materials from 3,000 gallons to 50 gallons. Hoffman-LaRoche went from 15,000 gallons of liquid ammonia to 2,000 gallons. In a decoupling example, Hoffman-LaRoche hired an outside contractor to make large-volume phosgene-based materials, thus eliminating the use of the extremely hazardous substance in a large plant where the collateral damage would be extensive. DuPont found a way to avoid keeping 40,000 to 50,000 pounds of deadly MIC, purchased from Union Carbide, in storage. It will produce MIC and use it immedi-

ately in a closed-loop operation, with a maximum of two pounds on premise at any one time. (Ashford et al. 1993, II-18)

As noted, chemical plants are expanding their presence in the United States. One thing we should do is to pay close attention to the location of new facilities. Adding them onto existing plants increases the risk of "collateral damage," putting both the new and old facilities at risk. New facilities should also be located far from rivers and harbors. The recent explosion in China makes the point. This occurred on November 2005 in the large city of Jilin, China. The plant, owned by CNPC, China's largest oil producer, is located right next to the Songhua River and just upstream from the city of Harbin, a city of 3.8 million people (roughly the size of Los Angeles). Chemical plants are frequently located next to such vectors, causing their accidents to have much farther-reaching consequences than is necessary, even to the point of being international in scope, as the Jilin accident demonstrates. The November 13 explosion released one hundred tons of benzene, nitrobenzene, and other toxins into the river, which is a primary source of drinking water for Harbin's citizens. Because benzene can cause leukemia and ulcers of the mouth, Chinese officials were forced to shutdown Harbin's drinking water supply for five days until the toxins passed downstream. However, they did not do this until eleven days after the spill, and even then said it was for "pipe maintenance." Unfortunately, the toxins did not dissipate rapidly and in December reached the Amur River in Russia, of which the Songhua is a tributary. Once the toxins reached the Russian town of Khabarovsk, officials there were forced to shut off drinking water to the town's 580,000 residents. Additionally, officials imposed a ban on fishing on the Amur that could last as long as two years. Thus, an accident that caused only one immediate death ultimately affected millions of people due to the plant's location next to a river. (Spaeth 2005)

## CONCLUSIONS

Facilities that contain large amounts of hazardous materials, such as chemical plants, water treatment plants, and railroad tank cars

constitute the best targets for terrorists. Nuclear power plants could create greater destruction, but it is not as easy to disable a spent storage pool as it would be to apply explosives to one of the hundreds of large tanks along Chemical Alley in New Jersey, setting off other explosions and fires just a few miles from Manhattan. It could be a drive-by shooting, and all the terrorists would have to do is to then escape upwind. Of course, radiological weapons and biological weapons can kill more people than the one-to-seven million that might be killed in a particularly successful attack upon a chemical plant or railroad train, but those attacks are much more difficult to stage. (Kosal 2006)

The chemical industry has made some progress in reducing concentrations of hazardous materials and in using less-toxic substances in its operations. But the instances of these are in the dozens rather than the thousands that we need. The industry has resisted mandatory inspections and standards and members of Congress, dependent on the industry for campaign financing funds, have allowed this resistance to be successful. Local communities all too often are unprepared for even small accidents and overwhelmed by substantial ones. There is no indication that DHS funds have improved our ability to respond. If anything, they have diverted funds from first responders to those concerned with terrorism. The industry should do more with fences and guards and surveillance cameras, but it is an almost hopeless task in the face of dedicated terrorists. As is evident from the oil disasters in the Katrina hurricane, some in the industry do not even follow obvious safe practices. This is a wealthy, highly profitable industry. We should expect more from it, but can't reasonably do so until we hear from Congress. Despite congressional ability to oversee the DHS, it has failed to give it "authority to require chemical facilities to assess their vulnerabilities and implement security measures," something the GAO says it requires. A January 2006 GAO report notes: "DHS has stated that its existing authorities do not permit it to effectively regulate the chemical industry, and that the Congress should enact federal requirements for chemical facilities." (GAO 2006)

# 7 Disastrous Concentration in the National Power Grid

---

ON AUGUST 14, 2003, LARGE portions of the Midwest and Northeast in the United States and Ontario in Canada experienced an electric power blackout. The outage affected an area with an estimated 50 million people and 61,800 megawatts of electric load in the states of Ohio, Michigan, Pennsylvania, New York, Vermont, Massachusetts, Connecticut, and New Jersey and the Canadian province of Ontario. The blackout began a few minutes after 4:00 p.m., eastern daylight time, and power was not restored for four days in some parts of the United States. Parts of Ontario suffered rolling blackouts for more than a week before full power was restored. Estimates of total costs in the United States range between $4 billion and $10 billion. (Agency 2003) The failure stranded millions of office workers and commuters from Pennsylvania to Boston and Toronto. It left Cleveland without water, shut down twenty-two nuclear plants (always a risky procedure), caused sixty fires, and in New York City alone, required eight hundred elevator rescues.

The details of the causes of the outage show the familiar string of interacting small errors. It was a hot summer day and demand was high, but that was not unusual. The device for determining the real-time state of the power system in the Midwest Independent

Service Operator (MISO) had to be disabled because a mismatch occurred. The device (a state estimator) was corrected, but the engineer failed to re-engage it on going to lunch. Normally, this would not be a problem for the duration of a lunch break, but it just so happened (the refrain of normal-accident theory) that forty-five minutes later a software program that sounds an alarm to indicate an untoward event at FirstEnergy began to malfunction. (Yes, this is the company that had the hole in the head of its Davis-Besse nuclear power plant.) It was supposed to give grid operators information on what was happening on FirstEnergy's part of the grid. Independently, perhaps, the server running the software program failed. Not to worry, there is a backup server. But the failure was of the sort that when the backup server came on, the program was in a restart mode, and under these conditions the software program failed. Moreover, it failed to indicate that it had failed, so the operators were unaware that it was not functioning properly.

Meanwhile, in the MISO office, the state estimator was restarted, after lunch, but again indicated a mismatch; no one knows why. ("Normal accidents" frequently have such mysteries.) At this point there was no untoward event that the program would warn about—but that unfortunately changed. Independently (normal accidents are all about independent failures whose consequences interact), faulty tree-trimming practices and the hot weather caused one of the FirstEnergy's lines to go down; as it had cut maintenance at its Davis-Besse plant, it also cut maintenance on its transmission lines. For complicated reasons only partly related to the failures, neither the MISO nor the company became aware of this line failure for some time. Finally, the MISO noticed and took the tripped line out of service, but FirstEnergy's failed program did not allow FirstEnergy to know of either the trip or that the line was taken out of service. Three more lines shorted out on trees because of FirstEnergy's cutbacks in maintenance, but the utility's computers showed no problems. FirstEnergy only became aware when its own power went out, a signal hard to miss, and had to switch to emergency power. By then, in just seven minutes of cascading failures, eight states and parts of Canada blacked out. (Funk, Murray and Diemer

2004; Outage 2003, 136–37) As normal-accident theory puts it, the system was tightly coupled, as well as interactively complex, hence the cascade of failures in eight states and Ontario province.

## INTRODUCTION

Our national power grid, the high-voltage system that links power generation and distribution, is the single most vulnerable system in our critical infrastructure, and this is the first reason we shall examine it. It has been attacked by natural forces, disabled by industrial accidents, attacked by domestic terrorists, and threatened by foreign ones. It need not be as vulnerable as it is at present. The second reason for examining it is that until recently, it was an example of a form of network structure that other systems in our critical infrastructure might learn from. The third reason is that it shows how market forces, inappropriately released and with faulty regulation, can reduce its reliability.

Electric power is obviously essential for our society; virtually everything important depends on it. Large-scale outages cost billions, and inevitably some people die. But power outages from bad weather, overheated transmission lines, equipment failures, or errors by the utilities are generally of short duration—a few days at most—and many essential services have backup generators. So we are not as threatened as we would be by the radiation of a nuclear plant or the poisons of a chemical plant. A two- to four-day shutdown might only take one to two dozen lives. But the economic consequences can be very large. Even without a large blackout, the costs of routine failures are substantial. One study puts the cost of interruptions and quality problems at more than $100 billion per year in the United States. (Priman n.d.) Most important, here is where terrorists could create havoc on a much larger scale than our big blackouts have caused. They could disable equipment that could take several months to replace and would have to be custom built in a place that had power, and a really large attack could leave a substantial part of the country without power—and without the

power to replace the power. In such cases the death rate would soar. I think such an attack is very unlikely, but so was 9/11.

Despite the disruptions and costs of the 2003 blackout, it is worth taking a quick look at another blackout to illustrate the resiliency of society to blackouts, or at least Canadian society. A large area of Canada, including the capital, Ottawa, was without power for weeks after three severe ice storms in short order in January 1998. In the bitter cold weather, sixty-six municipalities were without electric power, some for three weeks. A study of the response to that disaster found no panic and a smoothly working emergency-response effort. (It helped that there was ample firewood.) All the official response groups performed as trained; they had valuable experience from previous outages. Though this outage was extreme, they innovated and exchanged roles with other teams, and temporary, emergent organizations materialized to handle the inevitably unanticipated emergencies. Army crews cut firewood; private station wagons hauled in bulk food; the police, with time on their hands since there was no looting or crime, checked homes to find old people in danger of freezing; generators were flown in from distant provinces and trucks were used to power gasoline and diesel fuel pumps. There was endless innovation, and almost no deaths in three January weeks of extreme cold with no power. (Scanlon 2001) (The author of that study, Joseph Scanlon, also documents another successful Canadian disaster response when the Gander, Newfoundland, airport suddenly received thirty-eight diverted flights with 6,600 passengers at the time of the September 11, 2001, World Trade Center and Pentagon attack. The town of 10,387 grew overnight by 63 percent. Careful emergency planning and experience with diverted flights were the key to the amazingly flexible and successful response. [Scanlon 2001; Scanlon 2002])

Americans would not be as resilient if there were a terrorist attack on multiple sites of the U.S. grid. If one generating station were to fail because of an industrial accident, a storm, or even a terrorist attack, the consequences would not be great. The grid would find other sources to make up for the lost power. But if there are multiple failures (which is what we had in the Northeast 2003

blackout) the fragile web not only breaks in multiple places but brings down other stations in a cascade of failure. If the attacks not only disrupted power supply but also ruined generating equipment, much of which is custom made, it would be more serious.

I think the terrorist threat to the grid is very low, but if terrorists were able to electronically break into the control system of more than one large generating facility, their cyber attack could shut down a large portion of the national grid. Al Qaeda documents from 2002 suggest cyber attacks on the electrical grid were considered. We know there have been attempts by hackers, but probably not terrorists, to break into what are called the SCADA control systems (supervisory control and data acquisition) that run large plants. One utility reported hundreds of attacks a day, and a computer security expert said that there were a "few cases" where they have had an impact. Security consultants say that they were "able to penetrate real, running, live systems" and could disable equipment that would take months to replace. Terrorists could do this from another country without leaving a trace. (Blum 2005; GAO 2004b) They could also disrupt the national grid with a few well-placed small bombs at transmission towers. (A major concentrated line linking California and the Northwest is at the California-Oregon border.) Environmental extremists have blown up transmission towers in California. A large, well-organized terrorist group could attack all three of the major grids at the same time. Such an attack, or one on the operating systems, would do far more damage in 2006 then it would have in 1990, before the grid was deregulated and long-distance transfer of energy became a lucrative business proposition. The grid now is a larger, more attractive target. Its recent concentration can have disastrous consequences for our society.

## THE GRID

Our national power grid, made up of three independent grids that are loosely tied together, is an example of a highly decentralized system. It has grown steadily and rapidly as the population, urban

concentrations, and reliance on electricity have grown. Its structure is flat and slim, with a voluntary coordinating group formed in 1968, the National Electric Reliability Council at the head, presiding over ten councils. Below that, the structure and the nomenclature is changing yearly. Some say there are 150 electric control operators, others speak of 120 "cohesive electrical zones." Either way most of these are run by the dominant organizational form, investor-owned utilities—that is, for-profit utilities. In the works are regional transmission operators and independent service operators (ISOs)—there are five ISOs now—encouraged by federal legislation. These will be inserted between the ten councils and the 150 electric control operators to facilitate transmission and coordination.) At the bottom are more than three thousand local utilities and in addition, constituting 12 percent of the output in 1998, many other diverse organizations generating and sometimes selling electricity, mostly industrial plants. This network represents an enormous investment, including more than 15,000 generators in 10,000 power plants, and hundreds of thousands of miles of transmission lines and distribution networks, whose estimated worth is more than $800 billion. In 2000, transmission and distribution alone were valued at $358 billion. Considering the size of the grid, its structure is very flat.

The three grids are the Eastern Interconnected System, covering the eastern two-thirds of the United States and adjacent Canadian provinces; the Western Interconnected System, consisting primarily of the Southwest, areas west of the Rocky Mountains, and adjacent Canadian provinces; and the Texas Interconnected System, covering Texas and parts of Mexico.

It is hard to imagine what has been called the world's largest machine (a name also given to the Internet) as having only five hierarchical levels (resembling the Internet), with the top being a voluntary council and the bottom having more than three thousand units. But while there are central control rooms for each the three national grids, managed by the North American Electric Reliability Council, and control rooms for the 150 ISO coordinators, they would better be labeled *coordinating* rooms, since they act more as

traffic police and adjusters, responding to electronic signals that convey the state of the system. They rely on increasingly sophisticated automatic devices.

Massoud Amin, an authority who has written extensively on the prospects of self-organization through electronic "intelligent agents" on the grid, says it cannot have a centralized authority because of its size and complexity: "No single centralized entity can evaluate, monitor, and manage all the interactions in real-time." (Amin 2001) Demand decisions are, of course, made by the final consumers, as in any economic enterprise, but the decisions about balancing the system—how much power and from what generating stations will be put on which of several possible transmission lines to be sent to the distributor—are made by either the 150 ISOs, where they are in operation, or even much smaller facilities in the areas that are still being regulated. (Surprisingly, most of the system has not been fully deregulated. Many states in the Southeast have resisted, for example.) These "decisions," moreover, are increasingly not made by humans, though humans of course have designed the "agents" and the decision rules they will follow. Decisions are made by electromagnetic, and increasingly, electronic controls. A government report puts it as follows: "[T]hrough proper communications (metering and telemetry), the control center is constantly informed of generating plant output, transmission lines and ties to neighboring systems, and system conditions. A control center uses this information to ensure reliability by following reliability criteria and to maintain its interchange schedule with other control centers." (Administration 2000b) (That was in 2000; since then, information is being shared less because of increased competition.)

According to Amin, only some coordination (as of 2000) occurs without human intervention; much is still based on telephone calls between the utility control centers. (Amin 2001, 24) Indeed, when our transmission system as a whole is considered, we have what has been called a first-world country with a third-world grid, with much of the equipment for transmission dating from the 1950s. But automated coordination is a realistic goal, according to Amin.

(In an unpublished editorial submission, he deplores the response to the 2003 blackout, citing short-term fixes of problems that will only expand the scope of future disasters, and the failure to fund research for more sophisticated devices. [Personal communication, 2003]) In an enthusiastic 2000 article, he wrote:

> [I]ntelligent agents will represent all the individual components of the grid. Advanced sensors, actuators, and microprocessors, associated with the generators, transformers, buses, etc., will convert those grid components into intelligent robots, fixed in location, that both cooperate to ensure successful overall operation and act independently to ensure adequate individual performance. These agents will evolve, gradually adapting to their changing environment and improving their performance even as conditions change. For instance, a single bus will strive to stay within its voltage and power flow limits while still operating in the context of the voltages and flows imposed on it by the overall goals of the power system management and by the actions of other agents representing generators, loads, transformers, etc. (Amin 2000)

The automatic devices planned and being installed may be close to what has been called *cognitive agents,* or intelligent agents; that is, the agent *learns* from repeated experience. Just as voice-recognition programs grow in complexity and improve greatly as they encounter slightly different contexts, so do these agents.

Though I have not found an explicit discussion of this, it would follow that as new areas are added to the grid, they would continually organize themselves (the loose parallel to self-organizing systems), and also prompt the reconfiguration of the section of the grid they are attached to. While adjustments are made by operators, the relays, breakers, and multitude of other devices are not individually adjusted by hand, but through automatic devices. This kind of expansion, or growth dynamic, is not possible with a chemical plant, a hospital, or the CIA, no matter how decentralized they might try to be in the conventional sense. It does not mean that the utilities—say, Consolidated Edison or Pacific Gas

and Electric—are organized this way; these remain centralized bureaucracies. But the grids the utilities attach to appear to be these
radically decentralized, self-activating, almost self-organizing systems, run by professionals who are reasonably independent of the
utilities, though the utilities fund them.

It takes a very high level of technology to design and build the
billions of parts and thousands of intelligent agents that go into
the system, and to design coordinating rules. The system has
grown in an evolutionary way; no central authority planned it and
laid it down at one time. Many mistakes have been made along the
way and corrected, and "congestion"—where a transmission line
threatens to become overloaded—is a persistent problem with
rapid growth of the system. The system still has many reliability
flaws, but most result only in minor disruptions, generally associated with peak demands during hot, humid weather.

It is true that there have been very serious blackouts in the
United States, and normal-accident theory would say serious failures are to be expected because of interactive complexity and tight
coupling. Equipment failures in the transmission lines and unanticipated shutdowns of generating plants are to be expected, but
they may interact in unanticipated ways with each other and with
peak load demands, or with the phenomenon of reverse flow of
current because of "loops."[1] "Normal accidents"—simultaneous,
interacting, multiple failures that are unpredictable and even hard
to understand—are quite rare, as normal-accident theory would
expect, and remarkably so, given the excessive demands on the sys-

---

[1]Here is how a government report explains it: "The interconnection of the transmission grid makes management a difficult and challenging task. One of the
biggest problems is managing parallel path flow (also called loop flow). Parallel
path flow refers to the fact that electricity flows across an electrical path between
source and destination according to the laws of physics, meaning that some power
may flow over the lines of adjoining transmission systems inadvertently affecting
the ability of the other region to move power. This cross-over can create compensation disputes among the affected transmission owners. It also impacts system reliability if a parallel path flow overloads a transmission line and decisions
must be made to reduce (curtail) output from a particular generator or in a particular area. An RTO [Regional Transmission Organization] with access to region-wide information on transmission network conditions, with region-wide

tem and its size and complexity. Much more likely to cause failures are single-point failures, such as overload because of weather and demand. (I will deal later with deliberate destabilization to manipulate prices, as in the Enron case in California.) In the 1990s, demand in the United States increased 35 percent but capacity increased only 18%. (Amin 2001) One does not need a fancy theory to explain large failures under those conditions. (Indeed, one *does* need a fancy theory, such as network theory, to explain why there were so *few* failures under these conditions.)

Failures can be large: one failure in 1996 affected eleven U.S. states and two Canadian provinces, with an estimated cost of $1.5–$2 billion; the 2003 failure in the Northeast was even larger, at $4–$10 billion. But most are local. The appearance of deregulation in the second half of the 1990s took its toll. In the first half of that decade there were forty-one major outages affecting an average of 355,000 customers. That is about eight medium-sized cities each year, but in terms of the nation, very small. In the second half, as deregulation began to take hold, there were fifty-eight major outages, affecting 410,000 customers, a 41 percent increases in outages and 15 percent increase in the average number of customers affected by each one. The 2003 Northeast grid failure was our most severe, and we came within seconds of an equally large one in 2005. So the grid is becoming less secure.

But I am more struck by the high technical reliability of the power grids before deregulation than by the few serious cascading failures they have had over the decades. Without centralized control of the three vast grids, despite the production pressures of mounting demand and despite increased density and scope, they muddle through remarkably well when not being manipulated by energy wholesalers and starved of investment by the concentration of energy producers. What can we learn from this that might be of use in redesigning other parts of our critical infrastructure?

---

power scheduling authority, and with more efficient pricing of congestion can better manage parallel path flows and reduce the incidence of power curtailment." (Administration 2000a)

## LESSONS FROM THE GRID?

These principles seem to have been at work:

(1) It has had bottom-up growth. The network evolved gradually as needed for reliability. It was not imposed from the top, either by a few dominant for-profit firms or by a strong central government. (State and federal agencies played monitoring and regulating roles, entering the scene only to deal with evident problems.) Replicating this feature in other existing systems in our critical infrastructure will be difficult. Something other than evolution is needed to reduce the concentrations of hazardous materials or populations in risky areas.

(2) The network formed on the basis of independent units voluntarily coming together for reasons other than horizontal market control (though that has always been a danger and is a large one at present). Horizontal market control is when a few large entities in the same business coordinate practices (collude) to achieve above-market returns, or excessive profits. (Vertical market control is where one organization integrates key functions and sets excessive prices, without fear of competition.) The local utilities cooperated with each other in order to increase reliability. Since they were largely monopolies, they could cooperate without fear of price competition; local governmental bodies were only to ensure that the rate of return was fair. (Admittedly, a weakness of the system was that the utilities had undue influence over some local bodies, receiving above-normal rates of return, sometimes buried in deceptive accounting schemes.)

(3) Shared facilities (transmission lines in this case) are routinely accessed and priced at cost, or close to it. (With deregulation, the transmission firms must be competitively priced, turning a common good into a commodity.) This exists in much of our transportation system now and in the Internet; it could be explored for the gas and oil pipelines critical for our infrastructure, with the federal government setting fair rates of return and incentives for modernization. Much of our critical infrastructure could be defined in terms of a common good, subject to regulation, and not as vulner-

able to price competition or unregulated monopoly. Supreme Court rulings in recent decades have greatly narrowed the scope of common goods.

(4) Members support independent research and development facilities for technical improvements that are shared by all. R&D investments have declined since deregulation because the market does not value them. If firms were required to contribute to an R&D fund, this amount would be removed from stock price evaluations. Something similar could be done where there is private underinvestment in security and reliability in industries with concentrations of hazardous materials and in other parts of the critical infrastructure, including telecommunications.

(5) Oversight bodies (generally voluntary) establish broad regulations and accounting practices (labeled *commonality interdependency* in chapter 9). This enables firms to establish interoperability where facilities are shared. It is a curse of decentralized systems that interoperability is difficult to achieve. Strong oversight bodies are required to ensure independent invention and innovation that still provides interoperability. Still, the electric power industry has done reasonably well here, and the Internet has done very well, so it is possible in these two very critical parts of the critical infrastructure. Less oversight is needed in other parts, since interoperability is less essential.

While I would argue that these characteristics of a sound network fostered the reliability and increased efficiency needed for the key element of our critical infrastructure, it appears that the electric power deregulation that started in earnest with the 1992 legislation and was extended in 1999 legislation will change its character. While officially designed to increase competition, and thus lower prices, there are signs that it increases consolidation and thus market control by the largest firms, and this could lead to excess profits and less reliability. The price of electricity, controlling for fuel costs, was falling before 1992 but has risen steadily since then, though deregulation was to have lowered it. It is worth reviewing recent developments to see where the dangers might lie.

## DEREGULATORY DANGERS

Electric power in the past fifty years is a good example of the possibilities of networking, but it also illustrates the perils. Networks are less subject to predatory behavior than other economic structures such as hierarchies or multidivisional firms because units cannot accumulate commanding power, reciprocity is more evident, and there is an awareness of a common fate. Nevertheless, they are not immune to predators. In the 1960s through 1992, the utilities were quite independent, loosely regulated, and gradually linking up with each other to ease demand fluctuations and handle more temporary outages. The connections were through the transmission lines, of course, and as these connections grew in number and significance, technology evolved to routinize and stabilize the intermittent exchange of power and to meter and price the exchanges.

Think of a group of manufacturers, making, say, complicated machine tools. They manufacture many parts for the tools. One or the other might run short on a part, call up a nearby manufacturer, and purchase some if the latter had the capacity to increase production. Next week the favor might be returned, at a price of course. If such exchanges became frequent, regularized systems of inventory and exchange might be installed, and a network established. If the volume exchanges became significant, one firm might specialize in handling these exchanges, perhaps forming a separate organization to do so. This is what happened to the power grid, as exchanges needed to ensure reliability increased. Common standards evolved, regulated by councils and the government.

The local utilities were monopolies, justified as "natural monopolies" because it would be wasteful to have competing plants and transmission lines in the same locale, and they were considered a common good in that everyone used them. They were regulated primarily by their local authorities, with regulators setting maximum returns for the utilities, and the utilities kept excess generating capacity in reserve to meet surges in demands and temporary failures in production or transmission. (Today these local utilities

are still monopolies but are less often independently owned. They are being consolidated into holding companies and large owner-ship groups and thus are not subject to local community control.)

At this point, before deregulation, we had a large number of in-dependent units linked together to share power when needed, in a quite decentralized web, or network. It allowed new units to be es-tablished to accommodate growth in the market with only minor adjustments at the edges of the system. Technological advances made many of these exchanges fairly automatic, such that few supervisory levels needed to be added to the system; it remained quite "flat," and very reliable.

If we had remained that way, it would have been a model for a number of industries and services that form our critical infrastruc-ture. For example, the power grids and transportation systems have a common problem, congestion, which can disrupt the grids and disrupt the possibilities for rapid evacuation and rapid move-ment of emergency equipment in transportation systems. Adding transmission lines are difficult in urban areas (because of high land values and safety considerations), and adding more highways or airports is similarly difficult, even though in both cases it is the crowded urban areas that most need the increases. Both systems at-tempt to increase the capacity of their existing transmission facili-ties. What the power system had was separate, independent, local generation and transmission facilities called independent power producers (IPPs), along with (1) research and development agen-cies they supported with fees (principally the Electric Power Re-search Institute, with a substantial budget, responsive to the locals) and (2) coordinating agencies that set standards and policed its members, principally the national Electric Reliability Councils. In keeping with the best web designs, the organizations above the IPPs coordinate, police, and manage the system, rather than direct it.

The transportation system, in contrast, consists of multiple or-ganizational interests with considerable clout: vehicle manufactur-ers, highway construction companies, insurance companies, gov-ernments at the national through local level, and the police. Some

were concerned with only their part of the system. In the case of truck and auto transport, the automotive companies compete with one another and share few common interests with traffic management systems within and between cities. Though the Department of Transportation attempts to coordinate and standardize devices that would reduce congestion, neither the cities nor the auto manufacturers are responsive. Manufacturing is highly centralized in a few auto companies, giving them the power to resist decongestion devices in the vehicles and in road design, and neither cities nor users are able to insist on necessary changes in vehicle designs. Cities themselves are not coordinated by any regional or national agency, nor are there any major R&D facilities that serve the industry, only scattered programs, mostly in universities.

This shows the difference between a *decentralized* system, such as the power grid, especially before deregulation, and a *fragmented* system. Decentralized systems have coordinating bodies at the top and incentives to cooperate to maintain the system. Fragmented systems lack these.

In Japan and Europe, where the national governments are much stronger, intelligent transportation systems (ITS) are much more advanced; even Europe, with its many sovereign nations, manages to cooperate and coordinate better than the states within the United States. One of the oldest and simplest ITS devices, the "easy pass," allowing one to drive through a tollgate without stopping, and receive a bill later, is not standardized in the United States. Truck driver must have a dozen or more stickers or placards as they traverse the nation, and information on the vehicle and its contents, vital for safety, is generally not registered by the toll station. A much larger percentage of European and Japanese vehicles are equipped with driver assistance devices warning of dangers and showing alternative routes, often responding to sensors in the roadway (these are lacking in the United States), thus reducing congestion and accidents. The transportation systems of Europe, despite the handicaps of dealing with multiple nations and cultures and density, with ancient cities and no room to expand, appear to have managed to

combine centralized planning and coordination with local authority on route planning and local safety concerns. So has Japan.

## THE LINKS IN THE NETWORK

The electric power grid and the Internet are similar in that they are
both one big system. But transferring electricity is quite different
from transferring electronic signals. Distances on the Internet are
largely irrelevant. Signals from England to Spain will generally
pass through North America. The electric power grid in North
America is also one in big machine in that everything is or can be
connected, but the distance between the place where the power is
generated and where it is consumed makes a great deal of difference. There are problems with congestion on the Internet, especially
with the advent of large video and music files, but the Internet's
electronic impulses are, in a sense, simple. They have addresses
that determine where they will go, and the means to select the most
efficient routing. In contrast, when electricity is generated, it must
go into a transmission line (it cannot be stored, or held back, as
electronic impulses can) and it will flow through the most open
line, making the routing a complex matter all along the way. Sending electricity is more complex, and the complexity increases with
the length of the transmission and the complexity of the interconnections, neither of which is true of the Internet.

The burgeoning field of network analysis pays almost no attention to the links between nodes, but the difference between the Internet and other network forms is important. Packets of information
on the Internet can travel over diverse paths and be recombined
(because each has all the addresses), and the paths show incredible
diversity because of the connection of the nodes. Overloading on
the Internet just slows things down a bit; on the power grid it heats
things up and interferes with voltages and such things as reactive
power. On the grid, electricity on the transmission lines will seek
the shortest path that is left open by relays even if this results in

overloads, improper voltages or phase transitions, and other causes of failure. The links are crucial. (They are even more crucial in two other network forms we will discuss: networks of small firms and terrorist networks.) We should think of the grid as a large number of clusters of cells where the links between the clusters are best kept short to prevent instabilities in the current that is transmitted.

Changes in the generation and transmission at any point in the system will change loads on generators and transmission lines at every other point—often in ways not anticipated or easily controlled. Frequencies and phases of all power-generation units must remain synchronous within narrow limits, but the limits are even narrower on the grid, where a small shift of power flows can trip circuit breakers and trigger chain reactions. (Lerner 2003) (A relay with a faulty setting brought about the 1965 blackout.)

These conditions have led some engineers to say that the grid is vastly overextended because of longer transmission distances, and regulation should restrict the amount of "wheeling," which means transferring power from one point to another via a third party (a transmission company). Most engineers disagree, at least judging from the published literature (the published literature, of course, reflects the industry's point of view). Their argument is that with the proper incentives—the chance to increase profits—the industry will invest in more lines and better facilities, which will allow safe long-distance transmission, and they say the power loss from long-distance transmission is not large. In effect they say that if all the "intelligent agents" that Amin speaks of are inserted, the system will behave more like the Internet and be as risk free. Shortly we will find that other experts disagree.

## INDUSTRY CONSOLIDATION

Starting in the 1970s, deregulation of common carriers, such as air and truck transport, began, and in the 1992 Energy Policy Act this strategy was applied to electric power, on the grounds that there

was insufficient incentive to improve efficiency, that excess generating capacity was wasteful, and the cost of electricity varied widely between areas (generally states and regions). (The act also allowed the industry to make campaign contributions for the first time since the 1930s scandals. The interests of both Congress and the industry were thus "aligned.")

(I am drawing freely from a report by Eric Lerner [2003] of the American Institute of Physics, a critic of deregulation. See also the work of former utility executive John Casazza, who predicted the increased risk of blackouts because of deregulation and has become an increasingly virulent critic of the industry he left. [Casazza 1998; Casazza and Delea 2003] He is routinely, but not always convincingly, contradicted by industry spokesmen. My other main source is the government's Energy Information Administration, which publishes annual reviews of the industry and, of course, favors deregulation.)

One technical argument for deregulation was that peaks and failures could be accommodated by more extensive wheeling. The economic argument for transmitting power over longer distances was that the Northwest and the Southeast had cheap power and it should be sent to areas where power was expensive. Wheeling was already extensive and grew at an annual average rate of 8.3 percent between 1986 (well before the post-1992 reforms) and 1998. (Administration 2000a, 24) Further deregulation occurred in 1999, and wheeling was to take a steep rise after 2000.

Deregulation also allowed utilities to buy up one another, since control by local governmental was reduced, and deregulation required the unbundling of generation, transmission and distribution. Unbundling generation and transmission would allow more flexibility, though most utilities still both generate and transmit. The Energy Information Administration (EIA), a part of the U.S. Department of Energy gives the details: expensive production facilities were closed down, and technological advances meant that it was no longer necessary to build a 1,000-megawatt generating plant to exploit economies of scale; combined-cycle gas turbines are maxi-

mally efficient at 400 megawatts, and aero-derivative gas turbines can be efficient at 10 megawatts. (Administration 2000a, ix)

Predictably, and intentionally, mergers took place. From 1992 to 1999, the investor-owned utilities (IOUs) have been involved in thirty-five mergers, and an additional twelve were pending approval. The size of the IOUs correspondingly increased. The ten largest in 1992 held 36 percent of IOU-held generating capacity; this went to 54 percent by the end of 2000. In 1992 the twenty largest already owned 58 percent; this rose to 72 percent in 2000. (Administration 2000a, x) While there are more than three thousand utilities, most are small and inconsequential; 239 IOUs own more than three-quarters of the capacity. This is a substantial consolidation of the industry; consolidation would somehow increase competition, it was curiously believed. (It should have lowered prices, but it has not. Historically prices had been in a slow but steady decline for fifty years, but between 1993 and 2004 the retail price for all sectors rose by 9 percent, and for residential customers, 8 percent. Deregulation was supposed to reduce the differences between localities, such as states. It has not. Idaho's rates are still half those of Connecticut, and similar though somewhat smaller differences obtain between other adjacent states.) (Administration 2005)

## THE VIRTUES OF INDUSTRY CONSOLIDATION?

Rather than see the system up until the 1990s as resilient, reliable, and expandable, the EIA describes it as "balkanized." "To better support a competitive industry, the power transmission system is being reorganized from a balkanized system with many transmission system operators, to one where only a few organizations operate the system." (Administration 2000a, ix) (Some believe that the prediction of "only a few organizations operate" is ominous indeed, and it suggests the immense consolidation the industry leaders may be seeking.) The EIA report clearly believes that regulation has

hampered efficiency. It does not deny that regulation was necessary in the past, and cites the 1935 Public Utility Holding Company Act that put the Securities and Exchange Commission in charge of "policing the widespread abuses of the large holding companies" and "aimed at breaking up the unconstrained and excessively large trusts that then controlled the Nation's electric and gas distribution networks." (29)

But the EIA believes that act has outlived its usefulness because the economic institutions that protect customers and investors are all working. As evidence, it cites the changes in the last decades that have precluded the predatory behavior of the trusts of the past. It is a remarkable list, since every one of the "changes" was proved to be illusionary by the scandals of 2001 and 2002, and was foreseen even as the agency was writing its report :

- The development of an extensive disclosure system for all publicly held companies
- The increased competence and independence of accounting firms
- The development of accounting principles and auditing standards and the means to enforce them
- The increased sophistication and integrity of securities markets and securities professionals
- The increased power and ability of State regulators. (49)

Not a single one of these promises has proved viable.

This is not to say that the decentralized monopolies of the past fifty or so years did not need reform legislation and invigorated regulation. The rates varied greatly, with some states charging more than twice as much as others, and often without the justifications of federal hydroelectric power keeping the rates in the Northwest low and other contextual variations. Innovation in transmission was probably not aggressively pursued by comfortable monopolies that saw little need for it. But there was no great need for long-distance transmission and thus no need to unbundle long-distance transmission from local generation by allowing firms to emerge that specialized in transmission. Two things made the changes dangerous.

First, the physics of the grid make long-distance transmission wasteful because current is lost in the form of heat (there is disagreement on the significance of the loss), and more important, it raises the risk of uncontrolled interactions. The basic problem is a "collision between the physics of the system and the economic rules that now regulate it," says Lerner. (Lerner 2003) The vast system of electricity generation, transmission, and distribution that covers the United States and Canada is essentially a single machine in one respect. But everything is not effectively connected to everything else. Instead, clumps or cells of local activity are connected to the geographically adjacent cells—or at least that was the case before deregulation. The interconnections were there to insure against a sudden loss of power, so they were not in steady use.

After deregulation, the interconnections were heavily used. Engineers warned that increased long-distance trading of electric power would create dangerous levels of congestion on transmission lines where controllers did not expect them and could not deal with them. The problem would be compounded as independent power producers added new generating units at essentially random locations determined by low labor costs, lax local regulations, or tax incentives. If generators were added far from the main consuming areas, the total quantity of power flows would rapidly increase, overloading transmission lines. "The system was never designed to handle long-distance wheeling," noted Loren Toole, a transmission-system analyst at Los Alamos National Laboratory. (Lerner 2003)

A further problem was that the separation resulted in an inadequate amount of something called reactive power, which is current 90 degrees out of phase with the voltage. It is needed to maintain voltage, and longer-distance transmission increases the need for it. However, generating companies are the main producers of reactive power, and with the new rules, they do not benefit from it. Its production reduces the amount of salable power produced. So transmission companies, under the new rules, cannot require generating companies to produce enough reactive power to stabilize voltages and increase system stability. Lerner notes: "The net result of the

new rules was to more tightly couple the system physically and stress it closer to capacity, and at the same time, make control more diffuse and less coordinated—a prescription, engineers warned, for blackouts." (Lerner 2003)

## THE PROBLEMS OF CONSOLIDATION

These were some of the technical arguments for opposing the kind of deregulation Congress and the Clinton administration set in motion. But there were political ones as well. There was the risk that energy wholesalers, made possible by unbundling transmission and generation, would become speculators—gaming the system, as Enron, Dynergy, El Paso, and other firms did. Unbundling would require aggressive regulation to prevent predatory pricing, but regulation was not what business or the administration wanted. The separation of generation, transmission, and distribution allowed a fourth party to play a role: wholesalers such as Enron who reconciled the difference between the generator with the cheapest electricity and the distributor who would pay the highest cost. Looking for the best bargain, wholesalers were supposed to reduce price differentials, that is, perform an arbitrage role, which can be the role of the speculator. (There were no provisions for regulating these wholesalers. Unregulated arbitrage invites costly and inefficient manipulation, as the California episode made clear.)

The transmission part of the system became something of an orphan. Those producers that owned transmission lines were required to carry the power of competitors over their lines—the lines were defined as a common good, like telephone lines and cables in the Internet. But common goods that can be manipulated require regulation, and consumer groups and publicly owned utilities protested that without careful regulation and expansion, there could be technical problems and abuses. After four years of litigation, the Supreme Court upheld the new regulations on transmission and the separation of production and distribution in 2000.

Producers and distributors used the lines, but they did not want to build new ones as the population grew and demand even faster, and communities resisted adding lines. With more energy being wheeled, the lines became overloaded.

In March 2000, the warnings began to come true. Within a month of the Supreme Court decision requiring transmission lines to be open to all, thus suddenly increasing the economic value of long-distance wheeling on a grid not designed for it, electricity trading skyrocketed, as did stresses on the grid. The number of events where line loads were relieved by shifting power to other lines increased sixfold in just two months. The average hourly frequency deviations from sixty hertz (a measure of safety) went from 1.3 million in May 1999 to 4.9 in May 2000 and to an alarming 7.6 by January 2001. As predicted, the new trading had the effect of overstressing and destabilizing the grid.

Not only did energy companies run things to the limit of capacity, but they gamed the system, causing widespread blackouts and brownouts immediately after the March 2000 Supreme Court ruling. Federal investigations showed that employees of Enron, Dynergy, and other energy traders "knowingly and intentionally" filed transmission schedules designed to block competitors' access to the grid and to drive up prices by creating artificial shortages, all the while chortling over the phone about defrauding poor widows. In California, this behavior resulted in widespread blackouts, the doubling and tripling of retail rates, and eventual costs to ratepayers and taxpayers of more than $30 billion. (It also led to the recall of the governor, who was blamed instead of the energy companies.) In the more tightly regulated Eastern Interconnection, retail prices rose less dramatically. Nationally, the cost of electricity, excluding fuel costs, had increased by about 10 percent in just two years after 2000. (Lerner 2003) The vaunted savings from deregulation have never been realized.

After a pause following Enron's collapse in 2001 and a fall in electricity demand (partly due to recession and partly to weather), energy trading resumed its frenzy in 2002 and 2003. Although power generation in 2003 had increased only 3 percent above that

in 2000, generation by independent power producers, a rough measure of wholesale trading, doubled by 2004. System stress has soared, and with it, warnings by the Federal Energy Regulatory Commission and other groups. The stress was revealed in the August 14, 2003, blackout of the northeastern United States, to which we will turn shortly.

That is the minority view, critical of deregulation. The majority view emphasizes the role of the free market. First, if the transmission companies have enough of a profit incentive, they will invest in the technologies that will eliminate the kind of problems the critics refer to. One should be allowed to be skeptical about this. Shareholders in these companies are unlikely to take a long-run point of view, seeing their returns drop in order to improve reliability. It is in the interests of the companies to keep operating, of course, but a massive blackout only interrupts their income for a short time. A massive blackout will cause billions in losses, but not to the shareholders or the company. Only strict federal regulation is likely to force the companies to spend the kinds of money needed to make the vast grid highly reliable, and the 2005 energy act is quite toothless in this regard. Even a staunch advocate of deregulation notes this in his excellent analysis of the 2005 energy bill. (Eagle forthcoming)

### ENRON

The Enron story does not illustrate any key vulnerability to natural, industrial, or terrorist disasters, but it suggests we have another case of executive failure, widespread malfeasance at the top that brought prison sentences for Kenneth Lay, Enron's head (not served because of his untimely death), and Jeffrey Skilling, Andy Fastow, and other top Enron officers. It also illustrates two institutional conditions that increase the vulnerabilities of our electrical power system: the unfortunate consequences of ill-advised deregulation and lack of government concern with economic concentration in our critical infrastructure. A key advantage to deregulation, according to its supporters, is that it smoothes out market

fluctuations, thus reducing disparities and the inefficiencies of a market that does not properly "clear." Furthermore, "Transmission congestion . . . increases consumer costs by frequently denying low-cost transactions in favor of high-cost transactions," according to a government study. (Energy 2002) However, transmission congestion may be preferred by producers who can sell at higher costs, and this became apparent in the California scandal. (This section draw principally on the first popular book to examine Enron after the scandal broke, *Power Failure: The Inside Story of the Collapse of Enron,* by Mimi Swartz with Sherron Watkins [2003].)

For Enron, this started with natural gas. Instead of simply trading gas—that is, buying it with a short- or long-term contract from a producer and selling it to a distributor, it started making financial trades with suppliers and distributors based on the movement of the market for gas in general, not particular amounts or sources. It began dealing in energy derivatives, that is, swaps, options, futures, arbitrages (buying an item in a place where it is cheap and selling it to a place which will pay more), and so on. Government oversight regulated such transactions in the gas market through the Commodity Futures Trading Commission, since derivatives presented many opportunities for abuse. But in January 1993, just before the Clinton administration came into office and replaced the first Bush administration, the chairperson of the commission secured a vote that exempted energy derivatives and related swaps from government oversight. The exemption was to have historic consequences. The chairperson securing it was Wendy Gramm, wife of senator Phil Gramm (R-TX), who was a close friend of Kenneth Lay, the CEO of Enron, and a proponent of free markets. When the Clinton administration took over a few days after the vote, Mrs. Gramm left the Commission and joined Enron's board, earning a $50,000-a-year salary, stock options, and other cash benefits. (68) Freed of oversight and regulations, Enron began trading heavily in energy markets, exhibiting the dynamism that led to a glowing Harvard Business School case, and to *Fortune* and *Business Week* heaping praise on the "CEO of the year" or the "company of the year" in the late 1990s.

Enron's creative interpretation of a new accounting rule (Financial Accounting Standards no. 125) issued in June 1996, allowed it to effectively book all the profit streams expected from a power plant purchase over the next several years in just one year. By buying up plants each quarter and declaring on its balance sheet the profits that it *anticipated* over the next several years, it could show quarterly profits, even if the plant failed to produce the profits in succeeding years or even failed entirely. (136) It moved into the electric generation business, showing astounding growth and (questionable) profits. (It failed in its attempt at energy delivery and related services.) But the main profits came from trading, free of oversight as a result of the 1993 ruling. Enron traders inserted themselves between enough buyers and sellers of power to make electricity-trading revenues soar, even if gross margins were still negligible. The revenues looked good—it did a lot of business— though a careful look at the actual profit on the business would show it was minuscule. (140)

But something more attractive than revenues was possible. With a lot of power under contract, and an understanding of the key nodes on the grid that the power had to pass through if there were disturbances in the grid, they could sell their power at crisis prices, reaping huge profits. Initially, in June 1998, Mother Nature provided the disturbance. A severe heat wave hit the Midwest a month early, unexpectedly catching several power plants off-line, since they were getting serviced in preparation for the coming hot weather. Then storms took out some of the backup plants. Some small utilities were caught short and defaulted on supply agreements, and "power that normally sold for $20 to $40 per megawatt-hour under long-term contracts suddenly hit $500 per megawatt hour." Worse yet, the new free-market rules had attracted many small power marketers, with contracts to sell small amounts from small utilities to the big utilities. They began defaulting as well; they didn't have access to the power they were contracted to sell. The big utilities had to turn to the spot market, and now Enron, which owned many of the producers under the consolidation that deregulation had promoted, had much of the spot mar-

ket. They could wait until prices spiked and sell the power on the spot market. "Prices jumped as high as $7,500 per megawatt hour in a matter of minutes." One Enron trader made $60 million for the company in one day. Enron was awash in "a sea of cash" even as other energy traders, some quite large but perhaps not as aggressive, failed and left the business. (141) The year was 1998.

Early heat waves and storms were not new, but this crisis was new. Before deregulation, "utilities bought and sold electricity to one another at reasonable prices when such regional shortages appeared." (141) Now there were middlemen between producers and retailers, and the powerful ones were in the same organization that controlled many producers.

The next California crisis was not caused by weather disturbances but by traders, and it was more severe. In 2000 and 2001, in addition to huge price jumps, there were four days of rolling blackouts, and the state's largest utility, Pacific Gas and Electric, was forced into bankruptcy. (It is estimated that California's experiment with deregulation has cost the state more than $70 billion.) Enron, Senator Gramm, White House officials, respected academics such as economist Paul Joskow (he heads a powerful energy research center at MIT, at the time funded in part by Enron), and government bureaucrats all gave reasons for the crisis that were demonstrably not true. There was no extravagant usage by the state's citizens as they claimed—California ranked as the second-most efficient energy consumers in the nation. In one crisis month, July 2000, they were using less energy than in July 1999, and there was no blackout then. Demand for energy never exceeded the state's capacity. Power usage on blackout days was lower than in previous years. The state had added 170 new generation and co-generation facilities in the 1990s, giving it enough power to meet its growing needs. It was said that there were inadequate transmission facilities, which is probably true, but that also had been true in the years with no blackouts and high amounts of wheeling.

But there was something new about the transmission lines: traders sat on them and they could be squeezed. Deregulation had created middlemen who profited by buying cheap and selling dear,

but some of the middlemen were from companies like Enron and El Paso that also controlled the production of power. Swartz and Watkins tell a part of the story so well I will quote them. They mention Timothy Belden, an Enron star trader (who later pleaded guilty to some minor charges in return for cooperating with federal investigators). On May 24, 1999,

> Belden tried to send an enormous amount of power over some aged transmission lines—2,900 megawatts of power over a 15-megawatt path. Such a plan was destined to cause congestion on the line. California had an automated response to overloaded lines. Immediate electronic requests went to all the state's suppliers—*Do you have power coming across this line? Can you remove it? We will pay you to take it away!* But on this day, there also happened to be a human watching the wires for California's Independent System Operator, who couldn't imagine why someone would send so much power over such a small line. "That's what you wanted to do?" the dubious operator from the ISO asked when she called to check to be sure that the transaction had not been requested in error.
>
> When Belden replied in the affirmative—"Yeah. That's what we did."—she became even more incredulous. "Can I ask why?" (Swartz and Watkins 2003, 240)

Belden could not give an explanation, but agreed "it makes the eyes pop, doesn't it?" She said she would have to report the transaction to the power grid regulators because it seemed pointless, and Belden concurred. But from Enron's perspective it was not pointless; it drove the price of electricity up 70 percent that afternoon. Enron was fined $25,000 a year later after an investigation, but Enron had cleared $10 million that day, and California customers overpaid by around $5.5 million. "California's human-free automated system was completely dependent on the honesty of the power suppliers." (240)

Enron and others were happy to pay the trivial fines when they were occasionally leveled. Promised lower rates and better service from deregulation, California's wholesale electricity rates jumped

300 percent, and Enron, Reliant, El Paso, and Dynergy reaped huge profits. (241) Amin's "intelligent agents" can be programmed to commit fraud. Something other than a technical solution to the transmission problem is needed.

The famous cap that California's regulatory plan had put on prices the power suppliers could charge customers, as inadvisable as that may have been, was not the problem, because it did not interfere with gaming of the system once they had engineered volatility. When the state got the Federal Energy Regulatory Commission (FERC) to install price caps on the suppliers, the out-of-state suppliers such as Enron immediately vacated the scene. By December 2000, the lights were going out, and California asked FERC to lift the caps so that power could come in again, which it did. The power came back at even higher prices than before, even though there was actually no shortage. One Enron executive became worried about what they were doing and wrote an eight-page memo detailing the tactics being used: false congestion on power lines, such as Belden engineered, transferring power in and out of the state to avoid price caps, and charging for services the company never actually provided. This was about the same time that Texas senator Phil Gramm said in an interview with the *Los Angeles Times* that Californians suffering "the consequences of their own feckless policies," and vowed, "I intend to do everything in my power to require those who valued environmental extremism and interstate protectionism more than common sense and market freedom to solve their electricity crisis without short circuiting taxpayers in other states." (243) And renowned professor Paul Joskow, in a *New York Times* Op-Ed piece chided California for not building enough power plants and praised deregulation. (243) Deregulation had certainly helped one of the contributors to his MIT energy center, Enron.

Disclosures as early as the spring of 2002 made it clear that the professionals running the West Coast grid were ordered to curtail production, to make inefficient and unnecessary circular trades, and to misrepresent carrying capacity, using the wholesalers' market control to increase profits. See the several early news accounts

of deliberate destabilization by Enron in the California case. (Kahn 2002; Oppel 2002; Van Atta 2002: Berenson 2002)

## THE LITTLE ENGINE THAT COULD

The unraveling of Enron, Reliant, El Paso, and other energy companies might not have occurred were it not for a small public utility district thirty miles north of Seattle. The Snohomish district had signed a nine-year contract for power at four times the usual cost in January 2001, when, like areas in California, it was experiencing blackouts. When Enron collapsed and filed for bankruptcy protection Enron sued the district for $122,000 for canceling what it considered an illegitimate contract. Rather than collect $400 per customer to pay the fine to Enron, the Snohomish district searched for evidence of the illegal activity at Enron. Snohomish lawyers tracked down recordings seized by the FBI from Enron's western trading hub in Portland, Oregon. After a short legal tussle with the Justice Department, the utility was granted access in return for sharing the transcripts with law enforcement, since Justice had no intention of examining the evidence. The district hired a consulting firm, which recruited three people to pore over 2,800 hours of tapes. Their quotes made the national news when the district went to court with them. For example, one trader asks: "Do you know when you started overscheduling and making buckets of money on that?" Traders were shouting "Burn, baby, burn" when a forest fire threatened energy supplies. Traders joked about stealing money from California grandmothers and about the possibility of going to jail for their actions. One transcript does us the service of explaining "arbitrage." It goes:

> "He just f———s California," says one Enron employee. "He steals money from California to the tune of about a million."
>
> "Will you rephrase that?" asks a second employee.
>
> "OK, he, um, he arbitrages the California market to the tune of a million bucks or two a day," replies the first.

(Conversations were recorded because they have evidence of verbal contracts which the firms want to keep.)

It is interesting that the little utility district had scant cooperation from federal authorities. The utility went to search the Enron's warehouses in Dallas, and turned up evidence the government first claimed it never had and then said it did not have the resources to transcribe it. But the "little engine that could" did the whole job for about $130,000. (Peterson 2005; Egan 2005; Gonzales 2004)

FERC in particular did little to speed up the investigation, and indeed might be charged with suppressing it. The failure of many of our regulatory agencies has increased since the deregulation movement in the latter part of the twentieth century. By 2001, or perhaps earlier, FERC had transcripts of conversations between another power wholesaler, Williams Companies, and a power plant operator in California, showing the two conspiring to shut down a power plant for two weeks to boost electricity prices and Williams's profits. FERC kept the evidence under wraps for a year and cut a secret deal with Williams to refund California $8 million it obtained through the scam without admitting any guilt. (Leopold 2005) It was a nice deal for Williams; it precluded any criminal charges, or any publicity, and the company probably made much more than $8 million in the two weeks. However, Houston-based Reliant Energy was not so lucky; it faces millions in fines and prison time for four of its executives, as of December 2005.

It was not until March 2005 that the commission determined that Enron was engaging in illegal activity at the time it entered into exorbitant contracts with the west coast states. It was the first time the commission has acknowledged that the contracts were signed under fraudulent pretenses, though they had evidence that at least the Williams company had done so since 2001. It blamed the Federal Power Act, and rightly so; the deregulation legislation overlooked the obvious possibility of gaming the system. In a December 2005 report to Congress, the Commission noted that the legislation governing energy markets "did not address market manipulation and there was little in the way of penalty authority. If the express prohibition of market manipulation had been in place

then, it is very possible that it would have deterred market participants from manipulating the market because they would have known the serious consequences of their actions." (Commission 2005) The commission was required by Congress to investigate market manipulation and it concluded sixty-eight investigations. But the only civil remedies that could apply involved refunds and "disgorgement of unjust profits." By December 2005, it had facilitated settlements of more than $6.3 billion, others had not been settled and lawsuits are still pending. Despite the flaws of the 2005 energy bill, it does contain a provision that is supposed to forbid market manipulation. One wonders why such a provision was not in the law that established deregulation of electric power.

I should stress that this is not a "bad apple" case. The attack on the regulation of electric energy goes back at least to the time when Senator Lyndon Johnson accepted large campaign contributions from Texas energy companies in return for deposing a highly effective official that headed the principal federal energy agency of the time. (The dramatic story is well told in Robert Caro's *Master of the Senate* [2002].) Nor was Enron alone; there were other energy companies such as Dynergy, El Paso, Williams, and Reliant rigging the market. Nor will these be the last, judging from the weak regulatory provisions of the 2005 energy bill. Free markets only work when entry is easy, guaranteeing there will be many sellers to give consumers a variety of choices, and when collusion is difficult. This does not obtain in the energy market.

According to an article by Jason Leopold, the 2005 energy bill was shaped by Reliant Energy, even while it faced millions in fines for rigging the California market in 2001–2002. "House Energy Committee Chairman Joe Barton has for years pushed for legislation to create more competitive electricity markets. Reliant, whose executives enjoy close a relationship with Barton, played a role in shaping the energy bill, and worked side by side with Barton for years in shaping a portion of the energy bill that deals with competitive power markets." The cozy relationship between Congress and corporations is illustrated by the fact that Barton's staff includes two former Reliant executives, and until recently, Reliant's

lobbying team included two former Barton aides. Joe Allbaugh, until recently the director of FEMA, was active in energy policy in 2000 when California was suffering, receiving updates on the crisis from President Bush's economic adviser, Lawrence Lindsey, himself a former member of Enron's advisory board. Reliant Energy hired Allbaugh's wife as a lobbyist, and she received $20,000 for consulting work during the last three months of 2000 from Reliant, along with equal amounts from two other involved energy companies, TXU and Entergy (which we met in the previous chapter). She was consulting with them while her husband was getting updates from Lindsey, Bush's adviser. Reliant need not worry about the expensive lobbying fees. Its stock has increased nearly tenfold since California's crisis. Reliant employees showed their loyalty to House Energy Committee chairman Barton by contributing more than $35,000 to his political causes between 2001 and 2004. In addition, Barton holds about $15,000 in Reliant stock. (Leopold 2005)

From such networks of power, our energy policies will be made. Without campaign financing and lobbying reform, our electricity supply is not likely to be more secure.

## DEREGULATION AND THE 2003 BLACKOUT

After the 2003 blackout, extensive investigations by a joint U.S.-Canadian task force criticized the Ohio utility and virtually everything else about the system. It called for Congress to create tough mandatory reliability standards for electric utilities—with penalties for companies that violate them—to prevent massive power failures. The North American Electric Reliability Council (NERC) supervises this part of the grid, but its standards are voluntary, and it has been accused of being too cozy with its member utilities. However, the joint task force declined to consider whether deregulation played a role in the massive outage (though it called for tough new regulation, an implicit acknowledgment of deregulation's failure), postponing this inquiry for a future time. As of

2006, the inquiry into the role of deregulation had not taken place, though there have been some stormy conferences devoted to this question. As NERC is financed by the utilities who have done very well under deregulation, it is not a topic it is likely to explore on its own.

Predictably, industry participants said that the blackout resembled those of previous decades and was unrelated to deregulation, while critics of the industry argued that competition resulted in large cuts in professional staff, cuts in maintenance expenditures, and more extensive wheeling of electricity over inadequate and poorly maintained transmission lines. (Wald 2005a) Predictably, New York Senator Charles Schumer, a Democrat, said that NERC's recommendations were only "baby steps." The blackout should have been "a big wake-up call, but the commission only hit the snooze button," he said. (Funk, Murray, and Diemer 2004)

It is expected that NERC, made up of the utilities, will be given the job of establishing mandatory regulations regarding the maintenance and operation of the grid. It is very unlikely that it will restrict the long-distance transmission of electricity, which provides such handsome revenues for the wholesalers. Until the physics of the grid can be made compatible with this usage, such restriction is what is needed.

We can expect more serious outages. In fact, we came close to one in 2005. What Canadian engineers called a "rare glitch" occurred in May. Two apparently independent protection systems failed, and the cascading failures across Ontario came very close to extending into the United States as extensively as the August 2003 blackout. Predictably, the utility said that the short blackout proved that this system was safe. The blip was contained and handled very well, it said. Engineers at the utility (who incidentally happened to be on strike at the time), said that rather than being an encouraging sign, it was just the opposite, a warning of how vulnerable the system was. They pointed out that the system was only about eighteen seconds from causing a massive two-country blackout, and their analysis was confirmed by outside experts. There were four major swings before the system stabilized, according to re-

ports filed with the Northeast Power Coordinating Council. Had it been a hot day, an engineer noted, cascading failures would have been very likely. (Struck 2005)

Errors are inevitable in complex systems, and we have safety devices to prevent their spread. But this close call was closer than most. One of the two independent failures was a switching mistake by an operator after routine maintenance that caused a massive short-circuit in a line carrying 500,000 volts. There is a protection for such an expected error, but the circuit breaker took three times as long as it should have to operate—we are talking about milliseconds here—putting a massive strain on the system. The system survived, said a utility manager, but just by good luck. (Struck 2005) We can't count on luck forever.

### CONCLUSIONS

Electric power is vital to our nation. Increasingly, everything in our critical infrastructure depends on it. It is second only to concentrations of hazardous chemicals in its vulnerability to a terrorist attack. A sophisticated terrorist attack on our power system could include consequences such as the release of hazardous chemicals and much, much more. Industrial accidents and extreme weather can shut down the electric power grid for two, three, or four days; but terrorists could damage machinery that would take months to repair because much of the equipment is customized. The grid is vulnerable to heat waves, ice storms, floods and hurricanes, and to the inevitable equipment failures and employee mistakes that plague all complex, tightly coupled systems.

Paradoxically, the power grid is also one of four examples of a system design that offers promise for other systems in our critical infrastructure. Deregulation since the 1990s has altered the design considerably, but we can learn from many features still present in the power grid. The network evolved gradually in order to provide reliability. Independent generating stations voluntarily linked together and provided transmission facilities. The result was a de-

centralized system of transmission with independent control agents and increasingly sophisticated automatic control devices, such as the relays. Price competition did not exist, and rates and profits were controlled by local governments. Prices fell every decade. Transmission of power from one utility to the other was limited to balancing local demand; a power shortage at one utility could be made up by another utility at a reasonable price, since the favor might flow the other way at another time. Research and development was reasonably adequate, and utilities supported nonprofit technical institutions.

Deregulation and competitive pressures were introduced in order to increase efficiency and lower prices. Neither of these occurred; the mainstays of efficiency—maintenance and a professional work-force—have both declined, and after a long-term secular trend of lower prices, the cost of electric power has risen since deregulation. Reliability has arguably declined. While massive outages are fairly rare, so that we will need two or three decades under deregulation to estimate its reliability effects, our biggest one occurred in 2003 and one of a similar size almost occurred in 2005. In general, as a result of deregulation, more and more power is being pushed farther and farther over transmission lines that can barely handle it. The design of the deregulated system did not provide incentives for maintaining, improving, or adding transmission lines. With intense competition, the lines are no longer seen as a common resource that all must invest in and not misuse; investment comes out of profits. Line misuse has been widespread, as evidenced by the California crisis, and it may have contributed to the 2003 blackout.

Deregulation in the electric power industry was prompted by the same shift in ideology that led to deregulation of the airlines, trucking, and some other industries. But political interests and industry lobbying dominated the deregulatory effort. Unsavory practices by Texas Republicans can be related directly to the Enron scandal. With close ties to a House committee chairman, Reliant Energy participated in the drafting of the 2005 energy bill, even while under indictment for its illegal actions in California in 2000–2001.

Is it possible to have a grid that realizes the advantages of long-

distance transmission and still be reliable and secure? Assuming that the power loss of long-distance transmission can be reduced, we would need two things. First, there would have to be more investment in electronic devices, such as Amin and others are projecting, in order to remove the grid from its "third-world" status. The investment should come from the electric power industry, which has been quite reluctant to make it. Second, much more regulation is required to prevent gaming by power wholesalers. This is not a market with easy entry and little chance of collusion. These changes would improve reliability.

However, the changes would do nothing to decrease the size of the targets available to terrorists. Some of electric power generation is being deconcentrated as industries with excess power can sell it to power companies. But there are only small numbers of these, and the generating stations are miniscule. Concentration is increasing as more massive power stations are being built by ever larger comglomerations of power generators. Though I doubt that these targets are as attractive for terrorists as chemical plants and nuclear power plants, they are even less protected and their disruption could wreck large swaths of our economy for many months. What is unreliable is also insecure, and what is insecure is also unreliable. We are utterly dependent on electric power. Whether the threat is from nature, bad management, or dedicated terrorists, we should not tolerate the concentration of this most vital element of our critical infrastructure.

# 8 Concentration and Terror on the Internet

----------------------------------

IN THE 1990s, FOLLOWING the First Persian Gulf War, the United States engaged in almost daily bombing of targets in Iraq, in response to Iraq's failure to comply with United Nations Security Council resolutions and its interference with UN Special Commission inspectors. Early in 1998, the buildup of U.S. troops and material in friendly spots in the Middle East intensified, in preparation for Operation Desert Fox, a major three-day bombing campaign. In February 1998, the Department of Defense discovered that intruders had broken into numerous secure DOD computers. They had obtained "root access," which would allow them to steal information, alter information, or damage the DOD networks. They suspected that it was a case of "information warfare," with the Iraqi government behind the penetration. The attacks went on for almost a month. Finally they were able to trace the intrusions back to a Internet service provider (ISP) in the Persian Gulf region. President Clinton was briefed and both cyber countermeasures and "kinetic" (physical) ones were considered. Had the hackers stolen the bombing plans? How secure were our networks?

With the help of Israeli and other foreign law enforcement agencies, the department traced the intrusions to two California teenagers, assisted by an Israeli teenager. Internet signals "hop" all

around the world, and in this case, the Persian Gulf ISP was one of the hops between the teenage hackers in California and the Pentagon. (Vadis 2004, 102–3) We did not bomb the Persian Gulf ISP. This gives us an idea of the state of security on the Internet in 1998; it is not much better in 2006. Unauthorized access has been gained to nuclear power plants and other power stations, financial institutions, and intelligence agencies as well as the Defense Department.

## INTRODUCTION

Think of a safely designed highway, with exit lanes, wide curves, good lighting, and safe speed limits. The Internet is like that; by itself it is very reliable, like the highway. Now put on the highway cars that can go double the speed limit or roll over easily or explode if hit in the rear. Some are driven by people who are eating, talking on the phone, drunk or doped, or too young to have good judgment. Finally add in faulty regulation of vehicle manufacturers, poor licensing standards for drivers, and few highway patrol officers. This represents the devices that get us on the Internet. The Internet is a marvel; but some of the devices that allow us to use it threaten to bring it down. (The analogy cannot be pushed very far, unfortunately; we will see how the Internet, its access devices, and their interaction are far more complex.)

The Internet has been called the world's largest network, always on, with millions of transactions every hour. In itself, it is fantastically reliable and quite secure (though that is threatened). But the devices that get us on the Internet are prone to glitches and failures, and do not provide a great deal of security. Unless those devices provide secure transactions, the Internet presents the largest target for fraud and terrorism that we have. A terrorist can exploit faults in the operating system of a computer, in its software, or in the servers it is dependent on to gain control of a nuclear power plant, a chemical plant, or a city's water system if these systems are linked to a public network, which unfortunately they sometimes are. The terrorist could read the plans of the Department of Homeland

Security and the Department of Defense and alter them, or disable their systems. (Of less concern for this book, criminals can gain access to financial systems, including credit card agencies and banks.) This is partly because the operating systems used by 90 to 95 percent of those on the Internet comes from one source, Microsoft, and for years, because of its market dominance, Microsoft had no economic incentive to make its products highly secure or even highly reliable. While we will focus on security, the guru of computer fallibility, Peter Neumann reminds us that in systems, reliability and security are directly related. Events that can happen accidentally can be caused intentionally, and events caused intentionally can happen accidentally. (Neumann 1995, 126–28) We have to be concerned with both reliability and security, and sometimes they are so interdependent that it is hard to distinguish them.

The lack of security for machines on the Internet is beyond doubt, but as yet we have no public evidence that terrorists have used it to any great effect. (There is little *public* evidence that *thieves* have used it to great effect. Such evidence would greatly embarrass business, so it has not been made public. The estimates of yearly losses to business run in the billions.) As contrasted to the insecurity of Microsoft's operating system and software, its unreliability is more a matter of annoyance than of disaster, at least so far. However, a study by a National Academy of Sciences panel (still in draft form as of September 2006) anticipates disastrous consequences of software failures as critical systems become more and more dependent on software. Much of the software in critical systems is Microsoft's.

First we will examine the operating systems and servers that link to and run the Internet and the World Wide Web. This is where the potential for disaster is the greatest. (The Internet and the World Wide Web are not synonymous: the Internet connects computers through telephone wires, fiber-optic cables, radio signals, and satellites; the Web is a set of servers that are connected to the Internet. The Web stores documents that can be sent to other computers through the Internet upon request. However, we will sometimes treat them as the same since we use the Internet to access the doc-

uments on the Web.) Then we will examine the Internet itself, as the world's largest machine, and discover why the network itself (but not a computer on it) is still inherently very secure, even though what runs on it and holds classified documents and operates parts of our critical infrastructure can be very vulnerable. Then we will examine how commercial interests seek to centralize what is presently a vastly decentralized system with open access. A common carrier such as a highway could be privatized and centralized, creating the vulnerabilities we might expect from concentration.

## GETTING ON THE INTERNET

It is useful to initially consider two separate systems: your computer, and the system(s) to which your computer connects in order to send or receive e-mail or view Web pages on the Internet. An operating system such as Windows provides the *software* that runs the chips, relays, and so on in the computer—that is, the *hardware*. (The hardware itself is only rarely a security issue.) Microsoft's Windows or Apple's Mac OS X are commercial; you pay for these operating systems when you buy the machines or upgrade to a new version of the operating system. (You can also buy a machine without a Windows or Mac operating system and run Linux or variants of BSD Unix on it for free, but this is rarely done and requires considerable expertise.) In the language of telecommunications, your machine is the *client,* which is *served* by the second system (though any machine on the Internet can act as a client, server, both, or neither). The second system provides services to send and receive e-mail and interact with Web pages.

As a client, your operating system on your computer takes your e-mail, for example, and sends it, via a phone line, or cable, to your Internet service provider (ISP), such as America Online or Comcast, or your organization's ISP. The ISP is acting as a server. The software in the ISP routes it to another e-mail server that the recipient can connect to. Your message is broken up into packets; they may all follow quite different routes, but each packet has the

addresses of all the others so they can be combined at the recipient's ISP.

Your operating system, the first of the two systems, is open to penetration by hackers (like the teenagers in California and Israel we met earlier), "crackers" (malicious hackers), agents of foreign governments, competing business firms, thieves, and terrorists. That is why you are urged to set up firewalls, use passwords, avoid clicking on suspicious links in your e-mail or on the Web, and even use encryption. The ISPs are also vulnerable in themselves and these also have security protections. As we will see, it is possible to disable an ISP without even penetrating it; one can overload it by sending it millions of messages. But though this second system, the Internet, is the means of access to the first, the operating system, it is the first that presents the greatest reliability challenge and security challenge. Let us examine security first.

### INTERNET SECURITY

Threats to the Internet come in various ways with varying levels of disruption. First, the most common and most annoying are malware—the worms, viruses, Trojan horses, and backdoor access programs that can disrupt your machine and even erase the hard drive, where everything on your machine is stored. Once your machine has been compromised, it can be used to send these threats to other machines; your machine becomes a member of a "botnet," after ro*bots*. (To get very useful definitions and discussion of the technical terms in any of the chapters in this book, but especially this chapter, go to www.wikipedia.org, and for this chapter especially, to webopedia.org. These Web sites are two of the noncommercial joys of the World Wide Web.) You are not aware your computer is being used. Companies that place ads on, say, Google's search site, pay Google a few cents every time a computer clicks on that ad, and Google gives some of that money to the Web site that carries the ad. Through the botnet the hacker, with a Web site, can get many computers to click on the ad and will get a percentage of the fee the advertising company gets from Google.

Much more important for our purposes, botnets are used in another type of attack, a "distributed denial of service" attack, where specific sites are bombarded with so many messages that they are unable to send or receive messages. Theoretically, the whole Internet could be shut down with such attacks, but aside from the difficulty of pulling this off, why would a hacker want to eliminate his or her toy, or a terrorist his or her principal means of communication? Next, as we shall see, an attacker can take control of machines that are linked to the Internet and, for example, change settings on relays in a power plant. Finally, by gaining control of operating systems, the attacker can take over or modify data within a system. A law enforcement system could be gradually subverted in this way, the intelligence of intelligence agencies corrupted, or financial systems accessed and accounts changed or drained.

Everything connected to the Internet is vulnerable to these threats to varying degrees. We are all familiar with the first type of threat—worms and such—but it is the latter two types of threats that are the most consequential: taking control and modifying data. I will have much to say about the lack of security of Microsoft products in terms of the company's failure to make the most widely used operating system with its software invulnerable to worms, viruses, and the like. But when we come to the second vulnerability, taking control of systems and modifying data, the security failure has as much to do with the failure of the industry as a whole to protect itself as it does with the vulnerability of the industry's leading operating system.

The perpetrators of attacks come in three flavors. First there are the irresponsible hackers who are playing around, trying to penetrate protected systems to show that they can and to make trouble. Because of its ubiquity, or perhaps even more salient, its resented power, machines using Microsoft's Windows are a frequent target. When asked why he robbed banks, Willie Sutton replied, "that's where the money is." Windows is where 95 percent of the operating systems are. Some of these attackers are so malicious and do so much damage that they are called crackers. Second, there are criminals who steal money, and, rarely discussed, businesses and governments that require employees to engage in criminal acts of

espionage. (We should not be surprised that business and governments engage in espionage, and today the most productive venue for that is the Internet.) The third group are the terrorists who seek to disrupt the critical infrastructure of electric power, nuclear power, industrial plants, and the Internet itself in order to kill people or wreak economic harm. This group is presumably very tiny; indeed, we are not certain there are any terrorists attacking the Internet. The size of the second group is also unknown and presumed to be small in number but very effective; the victims are unwilling to have its extent of criminal activity publicized. The size of the first group seems to be enormous, but in fact may be a very small number of people; they are working in a system that has a huge amplification potential. If everything is linked, which is the genius of the Internet, everything is potentially vulnerable, and just one hacker's virus or whatever will have tremendous scope.

Regarding the first group, there are hourly instances of damage caused by malicious software such as viruses, worms, and Trojan horses. According to a computer security firm in London, global damage from malicious software inflicted as much as $107 billion in global economic damage in 2003. The "Slammer" worm alone was responsible for nearly $30 billion in damages in one week. (Geer et al. 2003) (With figures like these, why does business rely on such a costly mechanism as the Internet? Because the economic savings the Internet provides makes these figures trivial!) Regarding the second group, criminals, as noted we have no public information on the extent of criminal financial activity. Even more hidden from view are the activities of businesses and governments who steal information and secrets and perhaps disrupt their targets.

As far as one can tell from the public record, the third group, terrorists, have yet to wreak damage through the Internet, though they may be financing themselves through criminal financial activities such as getting into bank accounts and transferring money out. However, there are strong indications that terrorist groups have sought to gain access to consequential military and industrial and financial sites, much as the teenage hackers described in the opening to this chapter. According to a Government Accountabil-

ity Office report of March 2005, security consultants within the electric industry reported that hackers were targeting the U.S. electric power grid and had actually gained access to U.S. utilities' electronic control systems. But computer security specialists reported that only in a few cases had these intrusions "caused an impact." We do not know anything about the "few cases," but it is disturbing that there would be any. What can be done for fun could be done by a strategic adversary. The report said that the constant threat of intrusion has heightened concerns that electric power companies may not have adequately fortified their defenses against a potential catastrophic strike. (GAO 2005)

There is no indication as to the motives behind those reportedly seeking control of utilities' control systems; they could have been opportunists just demonstrating the vulnerability of the site or malicious hackers making trouble, which is worrisome enough, but they could have been terrorists exploring our vulnerability. The FBI reported in 2005 that there was evidence that terrorists were becoming more sophisticated in their Internet activity, but its illustration only concerned secret communications. For example, they were embedding messages in pictures that can only be read with a key (remember secret writing with lemon juice and then heating the paper to make it visible?). (It has a name: steganography.) But it did not see much activity. The FBI's top cyber-crime official said that terrorists are still unable to mount crippling Internet-based attacks against U.S. targets, and the agency has detected no plans to launch cyber attacks. However, a worm attack, presumably by a malicious hacker rather than a terrorist, almost shut down the FBI's Internet access. (Staff 2005b)

Widely used standardized technologies have commonly known vulnerabilities and one can go to Web sites to find programs that will exploit these vulnerabilities. It does not take a sophisticated hacker to launch malware. As a GAO report notes, "effective exploitation tools are widely available and relatively easy to use." Just type "hacking tools" into a Google search box and you will get millions of "hits," many of which offer instructions for hacking and directions to vulnerable sites. Hacking tools are especially

effective with the new "layered systems" features of the Internet that include peer-to-peer networks and chat groups. Social networks such as MySpace and Friendster, both of which have ten million or more users, can be attacked in still different ways. These layered systems are less secure and thus more promising targets.

Furthermore, it it is possible to identify from public sources the most heavily loaded transmission lines and the most critical substations in the power grid. A study notes that a hacker could change the settings on circuit breakers and at the same time raise settings on neighboring lines, and the diverted power would overload the lines and cause significant damage to transformers and other critical equipment, which might not be repaired for months. (GAO 2005)

Nuclear power plants are at risk simply from viruses. In January 2003, the Slammer worm that attacked the Microsoft server disabled a safety monitoring system for nearly five hours at our old friend, the Ohio Davis-Besse nuclear power plant. The plant's process computer also failed, and it was down six hours before it was available again. The virus also affected communications on the control networks of at least five other utilities. The utility claimed it was an isolated instance, but one may be skeptical as to how isolated it was. At least, the GAO report on the incident was skeptical about the utility's claim. It noted that there is no formalized process for collecting and analyzing information about control-system incidents, so there is no way of knowing how widespread such attacks were, and it called for strict reporting. (GAO 2005) No one has answered the call for strict reporting.

The lack of security can cause sizable economic damage, but it is rarely publicly documented. One such documentation, an indictment of a British computer administrator, is revealing not for the economic damage but for the threat to our national security. In November 2002, he was indicted on charges that he accessed and damaged ninety-eight computers in fourteen U.S. states between March 2001 and March 2002, causing some $900,000 in damage. More disturbing was that these networks belonged not just to private companies but to the heavily guarded Department of Defense

and the National Aeronautics and Space Administration. The indictment alleges that the attacker was able to gain administrative privileges on military computers, thus allowing him to copy password files and delete critical system files. The attacks rendered the networks of the Earle Naval Weapons Station in New Jersey and the Military District of Washington inoperable for an unspecified period of time. (GAO 2005)

In 1999, two Chinese military officers published a book promoting the use of unconventional measures, including the propagation of computer viruses, to counter the military power of the United States. In 2001, a loose coalition of Chinese hackers launched a successful denial-of-service attack on the CIA and White House Web sites, in response to the collision of a U.S. surveillance aircraft and a Chinese fighter jet. (Vadis 2004, 102) Actions by national states aside (and our Defense Department has the authority to launch cyber attacks on critical infrastructure sites of foreign states), it is not clear what breaking into military and other government establishments would do for terrorists, beyond being harassments, unless the establishments are government labs where hazmats are present. While it is true that they could disable disaster-response efforts after creating a disaster, it would require a very sophisticated terrorist organization to both create a large disaster and at the same time disable the agencies that are supposed to respond to it. The greatest threats to the Internet appear to be from hackers and crackers, but these are still serious threats. (Poorly designed software programs and applications may also cause Internet problems, but these will not be considered here. They are not strategic, i.e., intentional, threats as crackers and terrorists are, and though they may do damage there are no specific targets.)

It is worth emphasizing the interconnections in modern societies. Not only are all of our critical infrastructures connected to the Internet, but they are interconnected themselves. Thus an operating system failure or intentional penetration can interact in unexpected or even mysterious ways with seemingly unrelated parts of the infrastructure. A bug that causes a computer failure in a hospital pharmacy may result in the failure to deliver critical medi-

cines, patients getting the wrong medications, critical cases turned away to other hospitals, and reduction of surge capacity if there is also an unrelated crisis. Computer failures at airports sometimes ramify through the dense air transport network, creating economic hardships for business and government, as well as raising the threat of aircraft disasters. Errors are everywhere, and their unexpected interactions are facilitated by our interconnectedness, which is magnified by the Internet.

## DIVERSIFICATION AND REDUNDANCIES AS PROTECTIONS

In general, there are two ways to ensure the survivability of a threatened system. One is to have independent redundant components, such as a backup generator in a power plant to maintain it in a safe mode in case the plant itself fails to provide power. Backing up your computer files on a component that is independent of the computer, such as a CD-ROM or other storage device, is another example. Independent redundancy is the key to this form of survivability. Unfortunately, we have little of this in the operating systems that connect to the Internet. A virus designed to erase the hard drives on Microsoft operating systems could wipe out the data on all that are not protected. Even firewalls and other standard protections do not appear to be enough. The estimated $30 billion in damages that one worm caused in one week in 2003 affected sophisticated companies with firewalls and other protective devices. If 95 percent of computers running on the Internet have Microsoft operating systems, and this is the system the worm attacks, there is the theoretical potential of taking them all out by wiping the hard drive clean or freezing the computer so that only expensive and sophisticated recovery work would make them operable. It hasn't come anywhere near that as yet, and the vast majority of these machines are only used for e-mail and Web surfing. Furthermore, attacks on servers are more significant, and Microsoft does not have a monopoly here, but it makes enough of them

to do damage. And non-Microsoft servers can be vulnerable to worms attacking Windows machines. Worms such as Sober and Slammer can overload and damage the Net by repeatedly sending so many messages that these servers are overloaded and go down in a denial-of-service attack. In 2005, the third version of the Sober worm spread so quickly and widely that the FBI was bombarded with 200,000 e-mails a *minute* over *four* days, and it almost killed their system. (Staff 2005b) Adding computers that have the same vulnerability as the existing ones will not provide redundancy.

To protect from such cascading failure, we need the second form of protection—risk diversification, that is, diverse types of computers being used that have different operating systems and diverse programs from different sources. One virus or Trojan horse is very unlikely to affect them all. In a computer industry conference in 2003, a group of highly qualified engineers launched a broadside attack on Microsoft's "monoculture," in effect, its near-monopoly of operating systems. They drew an analogy to the farmer that plants several varieties of corn, since in any one year there could be a blight that kills one of the varieties but not the others. "This sort of diversification is widely accepted in almost every sector of society from finance to agriculture to telecommunications. In the broadest sense, economic diversification is as much the hallmark of free societies as monopoly is the hallmark of central planning." (Geer et al. 2003)

One could add that diversification is even more important where there are "strategic adversaries" such as hackers or terrorists. They could select the type of "blight" and the timing and method of delivery. Furthermore, over the Internet, the dispersion would be in microseconds. This makes the presence of a monoculture on the Internet even more serious than such modern monocultures as eBay, Amazon, and Wal-Mart. The possibility of another monoculture on the Internet is presented by the power of Google, but there are some reasonably competitive search engines from other suppliers, and search engines are not vital to our critical infrastructure. For the same reason, the near-monopoly power of Apple in the digital music realm is not critical.

The 2003 conference report created quite a storm. The report took the findings of the government's antitrust suit against Microsoft, which found extensive predatory behavior that led to Microsoft's monopoly position, and drew out the implications for *security* rather than economic competition. But the industry association that provided the venue for the report was not one that Microsoft chose to join; it included competitors of Microsoft. So another industry association to which Microsoft, but not its competitors, belonged attacked the report for "myopically looking to technology" (that is, Microsoft's technology) as the underlying cause of cyber breaches. It said that the root cause of most security problems is human error. "It's not this monoculture that is at fault here," a spokesperson said. "This started with human behavior, and it demands a human-behavior response." The association suggested that one of the best ways to protect against cyber intrusions is to increase security training for IT workers, that is, a human-behavior response. (Glasner 2003) Microsoft's products were apparently blameless, it was the users that should be blamed. Blaming the operator rather than the technology or system has a long history in analyses of failures. "Operator error" is the first and most common attribution when nuclear plants such as Three Mile Island or Chernobyl go awry, airplanes crash, and Bhopals happen. A more sophisticated analysis asks, "What in the system made it easy for operators or users to make mistakes?" (Perrow 1999; Reason 1990)

Why is the Microsoft operating system so vulnerable? An accepted principle of system design is to avoid complexity and tight integration of different components or interfaces. Complexity is the enemy of both reliability and security, engineers are fond of saying. The more complex the system, the greater the chance of the unexpected interaction of components, even when they are not faulty in themselves. A truly complex system may be impossible to fully comprehend, and the unexpected interactions may not appear for a very long time, and especially not in the short period of testing before marketing and installation.

Even if it were possible for all possible permutations of the system to be anticipated under all possible environmental conditions

and usages, there is the problem of failures. Everything is subject to failure; no computer program, no chip, or bus, or wire, or wire connection will always be perfect. Even if every component performs as desired, there may be unanticipated interactions. This, a result of complexity and tight coupling, is a second source of failure built into the design of the system. We guard against these two sources of failures—inevitable failures and those associated with system design—with redundancies, firewalls, alarms, bells and whistles because we know they will occur. But if there are multiple failures, even very small ones that each by themselves are inconsequential or are defended against, that interact in just the right, but unexpected, way, they can bring the system down, whereas the "single-point failure" is guarded against. The interaction can create a failure that no system designer could anticipate or comprehend, and no operator can understand. If the system is also highly integrated (that is, tightly coupled), there is a good chance that either the unexpected interaction of components or the unexpected interaction of failures in the component will produce a cascading failure. (This is the argument of normal-accident theory, previously referred to several times. Interactive complexity, in that theory, says that the interactions are not just linear like the stations on an assembly line but complex, with feedback loops and multiple uses. Tight coupling means that processes cannot be stopped or reversed, or substitutes or replacements added; they must proceed as designed. In a system with both characteristics, the accidents caused by interactive complexity and tight coupling are built into the design; while rare they are "normal." [Perrow 1999]) It is this possibility that the critics of Microsoft's "monoculture" raised.

But this just addresses the inevitable failures of interactive complexity and interacting failures. When we have a strategic attack, a deliberate attempt to disrupt the system or gain unauthorized control of it, complexity and coupling again play a role. Let's say a Microsoft designer adds a new feature, such as a Web browser or a music player, to an operating system such as Windows. Rather than making it an add-on or a plug-in, he integrates it tightly into the operating system, redesigning it to some extent. The designer

might claim that it is tightly integrated for efficiency reasons, but the federal court denied that claim when it was made by Microsoft. More likely, the integration was designed to achieve market share: the desired feature is difficult or impossible to uninstall so it becomes, by default, the most likely of its kind to be used, thus increasing market share. The unit becomes more complex and more tightly integrated. Complex interactions with the potential for failure occur when software is integrated with the operating system rather than offered as an add-on.

Tight integration means that a flaw that a malicious hacker or terrorist might discover and exploit will have much greater consequences than it would have in a loosely coupled or loosely integrated system. It will cause other features with which it is integrated to fail; for example, the device controlling pressures or temperatures in a plant can be inoperative. Or, the cascading failures occasioned by tight integration may allow unauthorized control or invasion by malware, as the GAO feared could happen to power plants.

The alternative is to use a modular design. The new feature can be plugged in or not, or a better version of the feature from another vendor can be substituted. (Modular designs are similar to the principles of networks we will examine in the last chapter. Networks of small firms, for example, are examples of the modular designs I am advocating here.) Microsoft integrated its version of an Internet browser (Internet Explorer) into its Windows operating system, claiming that it was tightly integrated and could not be removed, so customers buying Windows had to pay for it and could not ask for another browser instead. Those suing Microsoft for restraint of trade demonstrated that it could have been designed to make removal by Microsoft easy if the customer did not want it. Thus, the suit argued, the tight integration was not necessary but was deliberate. (The suit was successful but the remedy proposed— breaking up Microsoft—was overturned by an appeals court that required only minor changes in Microsoft's behavior.)

An example of Microsoft's tight integration is the SQL Server 2005, a relational database management system (not the same as

physical servers on the Internet). Instead of following the industry practice of selling components for data warehousing and business intelligence separately, Microsoft integrated them into its server. (Microsoft has many competitors in the data-warehousing field.) This resulted in a "radically lower price structure" but, in the eyes of some, created new dangers. A system architect commented as follows:

> "My concern lies in Microsoft's tendency to lock in the customer. I know that this is just how business works and Microsoft is not doing anything that other businesses aren't," he said. "What I do have a problem with is the direction in which such integration leads: potential difficulty to integrate a third party product into the mix and a software object of such size and complexity that change becomes more difficult and slower." "It seems that Microsoft stands alone in not understanding that tight integration of more and more features simply provides a greater vector for system failure, which is not easily fixed," he concluded. (Bekker 2005)

His comments are supported by the troubles Microsoft had with an earlier release of its server. One year after instituting its Trustworthy Computing program, it was hit by the Slammer worm, and its advice to "keep your system up-to-date with patches" was not even followed by Microsoft itself, since an internal company network was hit, even though a patch had been available for six months. The worm crashed servers and clogged the Internet, and disabled a nuclear power plant for hours, and it took Microsoft two days to get out from under it. The company acknowledged that "the patch management process is too complex." (Staff 2003) The failure of the new server brought down even more programs, such as a version of the SQL server packaged with Microsoft Visual Studio, and probably data-warehousing and business intelligence systems that were integrated into SQL Server 2005. The immediate problem was the software, but the tight integration propagated it. (However, the new Vista operating system, expected to be launched in 2007, is built with more small modules, making

development and testing easier to manage. Still, it is 40 percent larger than Windows XP, largely because Microsoft appropriately believes it must be compatible with previous releases.) (Lohr and Markoff 2006)

In addition to the problem of unexpected interactions stemming from sheer complexity and tight coupling, errors in design or software are more likely to appear in complex systems than simple ones. System designers or software programmers are less likely to conceive of the unexpected interactions that can occur. Furthermore, complex systems are more difficult to test. It is hard to conceive of all the possible environmental conditions the complex system might meet. The more complex the system—the more different and difficult things it tries to do—the more diverse parts of the organization it is embedded in will be exercised, and these must be tested as well. Individual modules that can be assembled to produce the desired effect are easier to inspect for errors and easier to test. But a modular-based system invites the production of modules by competitors.

Microsoft has favored complex and tightly-coupled products, rather than redesigned systems that are simpler and more reliable and secure, for a number of reasons. First, in the initial years of Microsoft's Windows program, it was used as a stand-alone operating system in the company workstations or for home computers. But once the Internet took off, everything became interconnected, and the unreliability of the workstation in a bank that had not been connected to the vast financial markets, now spread, as did the consequences of security breaches. Windows-based software is now critical for our financial structure, nuclear plants, industry control systems, defense department data, naval ships, aircraft, and so on. (Most everything that can kill is run by software, though not necessarily Windows-based: heart pumps, infusion pumps, pacemakers, medication distributions in hospitals; military weapons and fighting platforms; and now the safety of our most widely used weapon, the automobile.)

A second reason Microsoft favors complex and tightly coupled products is that redesigning the system in order to have the new feature without making the system more complex is harder than

just changing the design enough to incorporate the new feature. Simple designs for products that do complex things are, perversely, much more difficult to produce than complicated designs. It is necessary to move fast in a rapidly developing marketplace, while basic redesigns that reduce complexity take time. The alternative, providing modular features for plug-ins, is risky. A competitor might provide a superior plug-in, for example, or customers might resist buying the plug-in. Complex, integrated systems promote market control.

But it would be unreasonable to have expected a for-profit company such as Microsoft to have increased their difficulty of design and production, delayed introducing new features when competitors might get ahead of them, and invited competitors to plug into their Windows operating system and sell features Microsoft could otherwise sell or had to provide for free. The firm would lose its market dominance and thus profits. The cost of such tactics— some unreliability and a substantial loss of security for users—has been quite small for the company. Market share rose, as did profits. Even its illegal predatory behavior against competitors cost it only a little in fines and agreements to desist.

The security problem extends to the servers that match addresses to destinations and show the quickest route. Competition in the server field is substantial, and here reliability (and perhaps security, but that is not clear) is a key competitive factor. This competition and emphasis on reliability may have a great deal to do with the reliability of the Internet. The initial servers were built by the government and nonprofit organizations such as universities, and reliability and security were key goals. The introduction of private, for-profit firms into the server market occurred in the late 1990s, and they may have had to put reliability and security as top goals to compete with existing servers.

## COPING WITH MONOPOLY POWER

Since the beginnings of capitalism in the United States, customers have tried to force vendors to pay for the costs of faulty products,

but even in our litigious society where customers are free to sue, the courts have generally ruled that the buyer must beware. We have seen how little the fines are when chemical companies break laws and have accidents that kill and pollute; how little it cost the electric power companies (two to three day's revenue) to avoid maintenance, upgrading, and staffing expenses, but the cost to the rest of society runs five to ten billion dollars in an outage. But unfortunately, the reliability and security failures of Windows affects critical parts of our infrastructure, making us more vulnerable to industrial and terrorist disasters. We are not dealing with a product such as television sets, where product safety and security are barely an issue. Many critical parts of our infrastructure had nowhere else to go to get the immense advantages of computers and the Internet other than to Windows. Windows is not simply a consumer product, used for spreadsheets, e-mail, and Web surfing. It is found in all organizations in our critical infrastructure, and its vulnerabilities are frequently not walled off from the critical functions. It is just too valuable!

Even the U.S. Air Force does not have the resources to build a completely new, bug-free operating system, a word-processing program, graphic interfaces, and so on, so it has to use Microsoft products. But these are not reliable enough for the stringent demands of military aircraft. So the Air Force contract specifies that the software be configured in a more secure state, changing the system settings. It is able to review the highly secret, proprietary source code (but not to change it). (Theoretically, critical military devices are not connected to the public Internet, and thus are not accessible by hackers or terrorists, nor are the SCADA control systems of nuclear, chemical, and industrial plants, but in practice there are a variety of ways in which they inadvertently get connected or have to be connected. Simply accessing the SCADA system from computers in the head office of a power plant that happen to be linked to the World Wide Web will connect them, and we saw at the beginning of this chapter how hackers were able to get into the Defense Department computers and gain root access.)

The same problem confronted the Department of Homeland Se-

curity when it sought to establish a common computing environment for its twenty-two formerly independent agencies. It signed a
$90 million enterprise software deal with Microsoft for server and
desktop software for approximately 140,000 users. (Ironically,
three weeks later a critical security flaw was discovered affecting
nearly every version of the Windows operating system, including
Windows Server 2003.) A security expert, the former chief of staff
of the President's Critical Infrastructure Protection Board, said,
"The real alternative was to go open-source. But for 22 agencies,
an overwhelming majority of which use nothing but Microsoft operating systems, to convert to another platform in an efficient and
cost-effective manner would have been hard to accomplish. DHS
has neither the time, the money, nor the flexibility for that. Now it
is held hostage to the imperfections of Microsoft code-writing."
(Verton 2003)

Not only can Microsoft's products not be reengineering by users
to make them more reliable without having the proprietary code,
but its near-monopoly power prevents alternative systems from
appearing or surviving. But suppose users were spread out over a
half dozen different operating systems, and a half dozen word-
processing programs and other software attachments. A virus could
not disable them all, and disabling one of the six is unlikely to bring
down the other five. For competitive reasons, the six systems would
have to provide interoperability, just as Microsoft had to make its
bit-heavy, clunky Word program for word processing somewhat
compatible with the superior WordPerfect program of a competitor.
A critical facility such as a nuclear power plant could achieve true
redundancy by having three different operating systems and sets of
programs available in case one was disabled. It is also quite likely
that the six manufacturers would compete on the basis of reliability and security, and not just on the variety of applications, and
those most needing security would migrate to one of them. All six
would, in time, have high and comparable levels of security in order
to remain competitive. But with 95 percent of the market and Windows running on 330 million personal computers worldwide, Microsoft has little incentive to emphasize security and reliability.

Though the military is content to receive modified Windows programs to make it more safe and secure, the Commonwealth of Massachusetts has taken a different tack. In 2005, it standardized desktop applications on OpenDocument, an open-source format that is not supported by Microsoft Office. About 50,000 desktop PCs will be required at least to save their documents in the open-source format, and while they can continue to use their current version of Microsoft Office, they cannot purchase Office 12, which is due to be released in 2006. OpenDocument-based products will cost the state about $5 million, in contrast to the $50 million for Office 12. Microsoft could still compete with others for a program that operates on this open-source platform, but only if it supported the OpenDocument format, which it has so far refused to do. If other states adopted this tactic, and especially if the enormous U.S. government agencies did so, Microsoft's monopoly of this biggest part of the operating system software would be broken. In Massachusetts, a department might choose to use word-processing products such as OpenOffice and variants of it from companies including Sun Microsystems, IBM, and Novell. (LaMonica 2005) It is an example of how large customers could bring about deconcentration in an industry.

Microsoft is doing more in the security area today than it has in the past, but this has largely been in the form of patches to correct the flaws—and it appears that each patch becomes a target for a hacker, and the vulnerability returns. Programmers argue that making Windows more secure would require a wholesale change that builds in security from the beginning through rigorous testing and verification. But this could have the consequence of making 95 percent of the computers in the world gradually obsolete. Finally, hackers concentrate on Microsoft because it is the dominant system. If there were four, five, or six systems sharing the market, their efforts would have to be spent on four, five, or six systems and thus would be less intense on any one.

Competition among browsers may promote security. Users of Microsoft's Internet Explorer can be as much as twenty-one times more likely to end up with a spyware-infected PC than people who

go online with Mozilla's Firefox browser, according to a study in early 2006. (Kelzer 2006) There is also a vast difference between those two browsers in the time that they remain unpatched after a vulnerability is found. For 54 percent of the time in 2004, a worm or virus "in the wild" exploited one of the unpatched vulnerabilities in Internet Explorer. The comparable figure for Mozilla's Firefox browser was only 15 percent. (Krebs 2006)

Compared with Microsoft, the other operating systems, such as Linux or Apple's Macintosh system appear to be more reliable (though the evidence is mostly anecdotal), but not necessarily more secure. Security is a more recent demand than reliability, and perhaps the other systems, being less widespread, are not used in settings where security is paramount. There is still no great commercial advantage, as distinct from terrorist prevention, to offering secure systems, despite the reportedly enormous costs of thefts, and there may not be until there is a spectacular catastrophe or the government becomes sufficiently fearful of the damage that malicious hackers or terrorists can do.

One thing the government could do would be to make the suppliers of operating systems and software liable for security failures. This would happen if the courts ruled in favor of plaintiffs and if the Justice Department and its head, the attorney general, supported the courts' findings as they were reviewed by higher courts on up to the Supreme Court. This is not likely to happen. Both the Clinton and the Bush administrations have consistently rejected the notion of regulating vendors or users. The Clinton administration's information security plan stated that the president and Congress "cannot and should not dictate solutions for private-sector systems." The Bush administration's cyber-security plan of 2003 states that "government regulation will not become a primary means of securing cyberspace" and that "the market itself is expected to provide the major impetus to improve cyber security." (Vadis 2004, 112) We are not likely to get liability for software or operating system failures anytime soon, even if the issue is national security. However, findings of liability are starting to appear in a small way at the state and local level. If it grew, it would catch the

attention of all producers but would also impose very significant costs that would probably be passed through to consumers rather than reducing the large profit margins of producers such as Microsoft.

A draft paper by two members of the AEI-Brookings Joint Center for Regulatory Studies offers some excellent suggestions regarding mandatory reporting of vulnerabilities and liability for them. A recent California law requires mandatory incident reporting, and modeled on this, they suggest a federal program that did not release information about individual victims but reported only aggregated data. Of course these economists point out the drawbacks, principally that "reporting informs consumers, businesses, *and* malicious hackers" of the vulnerability. (Hahn and Layne-Farrar 2006, 56) Regulating cyber weapons is feasible from a legal standpoint, since there are already laws banning worms, viruses, cell phone scanners, and software that circumvents copyright protection. (57) Since the federal government is a key buyer of security software, consuming around 42 percent of all software and computing services as measured by revenue, the government could create a market for secure and reliable software if it insisted on it. The Department of Defense already requires all software to be tested for security by the National Security Agency. But when government agencies were rated for their overall information technology security, over half received either a D or an F. The Department of Homeland Security, which has a division devoted to monitoring cyber security, received an F. So did the Justice Department, which is responsible for investigating and prosecuting cyber crime. (60)

Insurance, the authors point out, could bridge federal policy and private-sector initiatives. The federal government requires all automobile drivers to obtain insurance before they drive, and the insurance is privately supplied. Cyber insurance is emerging as a viable option, the authors hold, and the insurance companies charge according to the degree of protection the company installs. "This route has the advantage of being market driven, so the price of additional security will be weighed against the benefits that the added security has to offer." (62) There are two rating firms at present

that act as third-party evaluators of a company's security efforts, but they apparently have little business. This would change if cyber-security policies were mandated, just as driver's insurance is.

## OPEN SOURCE

Is open-source software any more secure than proprietary software? A debate about this has been raging for years. Microsoft argues that its system is just as secure as those that use open-source software, such as Linux and Apache; others dispute that and add that, in any case, open systems fix flaws much more quickly. One security analyst argues that the "tight integration" found in Windows is not found in Linux, and this integration increases the number of security exposures in Windows. Furthermore, open and closed systems handle flaws differently. Most reports of flaws in Windows come from antivirus firms or from hackers. Exploitation of the flaws occurs "in the wild" (on the Net) and then we depend on commercial antivirus updates, followed by an operating system patch that Microsoft sends out. (Bill Gates is noted for blaming the victim, saying that it is the responsibility of users to keep up to date with the patches, rather than saying that it is Microsoft's responsibility to avoid the necessity of patches.) In contrast, university researchers or developers within the open-source community report security flaws more frequently, before they escape into the wild. These sources often provide immediate means to correct the flaw with an "emergency patch," and then, to take the case of Linux, a permanent change is approved by the inner circle of Linux code writers at the Open Source Development Labs. Open-source developers argue that their system corrects errors more quickly and can use new and more secure techniques. Because they are performing for their peers (other programmers they are directly or indirectly in contact with), they derive intrinsic satisfaction from these efforts. (Staff 2005g)

Some idea of the growing open-source enterprise can be gleaned from the statement of the OpenSSL Project, one of many such

projects. It says that the project is a "collaborative effort to develop a robust, commercial grade, full featured" toolkit with various protocols, including a cryptography library that is managed by a worldwide community of volunteers who planned, developed and managed the toolkit and its documentation. The toolkit is licensed in a way that anyone is free to use it for commercial or non-commercial purposes. The project "is volunteer-driven. We do not have any specific requirements for volunteers other than a strong willingness to really contribute while following the project's goal." (OpenSSL 2005)

The open-source movement has grown steadily over the years, along with systems such as Linux and Unix that utilize it, to the extent that it may be a threat to the business model of Microsoft. Anything that provides an alternative should improve security; users that require security can make a choice among systems on these grounds. (For a fascinating discussion of the operating superiority and security of the "freeware" Linux software versus Microsoft, see Kuwabara 2003.)

Indeed, Microsoft's dominance of the operating-system market and software products such as Office and its Word program may not last long. People are talking about "thin client" systems where a fairly simple device connects users to the Internet and to a virtual operating system on the Internet itself. For a small fee, users could rent much of what is now a desktop computer from the Internet and rent whatever programs they desire—the module option—which are then installed on the virtual computer. Or, with advertising, there may be no rental fee. These innovations may be shaking up Microsoft; the "Internet services model" is on its way with start-ups and established companies such as IBM and Google offering free or low-cost programs that do word processing, spreadsheets, and other basic functions of the desktop computer.

After losing top executives to Google and other companies, Microsoft bought a company in 2005 and thus acquired a brilliant programmer, Ray Ozzie, who has a distinct "unbundling" program in mind. The company could then offer online versions of its Office products, supported by advertising or fees.

It may be important that Ozzie, Microsoft's chief technical officer is seconding the criticisms I have been reporting and sounding the programmer's mantra. "Complexity kills," he wrote in a famous memo. "It sucks the life out of developers, it makes products difficult to plan, build and test, it introduces security challenges, and it causes end-user and administrator frustration." He even sees "Internet services" as offering a more open, competitive model. Steve Lohr, the *New York Times* journalist, says that Ozzie "speaks of a thriving 'ecosystem' of open competition in which developers and customers have many choices and in which Microsoft's future is not in crushing rivals but in becoming an attractive choice." Even open source is embraced. "I consider open-source software to be part of the environment, like the Internet," the Microsoft chief technical officer said. "It's not the enemy and it's not going to go away. It's great for developers." (Lohr 2005)

The issue of market domination, and its implications for security from both industrial and terrorist disasters, may not go away even if Microsoft loses its dominance and accepts "many choices," although that would certainly help. But we should be cautious. A company like Google, which continues to dominate the Web search market, with its extraordinary financial returns, may be able to launch a virtual computer with integrated features that is so successful that it dominates the market the way Microsoft now does. It then might not have to pay attention to security. Before this happens, it would be wise to make security flaws subject to liability suits.

Or, we could break up Microsoft into competing firms. Thomas Penfield Jackson, the judge in the Microsoft antitrust suit recommended breaking up Microsoft into two companies: a Windows operating system and an applications software firm. (This was rejected by an appeals court.) But this would still leave Microsoft's Windows monopoly intact. A group of very prominent economists recommended Microsoft be split into three Windows operating systems companies, and a fourth would be the software applications firm. This would introduce immediate price competition in operating systems, and the three mini-Windows companies would

have very strong incentives to keep their products compatible with one another while competing on reliability and security. Breaking up Microsoft would be very tricky, but decades ago, the Bell system was broken up by the government and the results have been very positive. (Miller 2000)

## THE INTERNET

Now we will leave the "cars" and turn to the "highway." The world's largest network is totally unique in organizational terms. It runs virtually without a central authority governing it; it is based on rules, or protocols, that are voluntarily accepted and open to modification; and while initially financed and designed by the U.S. government, it is now a mixture of public and private organizations, including foreign ones, with a minimum of central direction. Our interest in it is twofold: First, it is one example (we shall consider three others in the concluding chapter) of the possibility of a vast, decentralized, efficient, and reliable system that could be a model for some of the other systems in our critical infrastructure. Second, it illustrates the vulnerability of such systems to the economic dynamics of free-market capitalism, which intends to centralize a decentralized system. These dynamics can make it more vulnerable to errors and deliberate attacks.

Before the Internet came along (a primitive version was established a half century ago in the late 1960s), we relied on "dedicated" communication systems, such as the postal system, the telephone, and a few specialized devices such as stock tickers. These go from point to point. Dedicated systems are cheap and efficient. But the inventors of the Internet, working out of the Defense Department, chose a more complicated interoperability system. Instead of limiting a computer at the University of Chicago to make only one connection via a telephone line to a computer at Los Alamos, the computer could connect to a network that might have several computers on it. Furthermore the connection could allow a variety of transactions, such as file sharing, e-mail, or one com-

puter taking command of another. Establishing a platform from which more than one program could be launched of course made it more complex, something we should beware of. But it made the Internet open to innovations and new uses, including Web pages and now video and telephony, which the original inventors did not conceive of. In contrast, AT&T fought vigorously to control its monopoly over the dedicated telephone system when Carterphone and other innovative devices were proposed to run on its platform.

Fortunately, AT&T lost that fight. In 1969, in the Carterphone decision, the FCC ruled that the company could not discriminate against innovators that wanted to improve the phone service; in time, this decision caused AT&T's monopoly position to erode, and we had an explosion of innovations. When the Internet came along, the FCC again ruled that access to the telephone wires and the fiber-optic cables had to be open to all, for a reasonable fee. They were defined as common carriers, such as highways or rivers. This led to the creation of thousands of Internet service providers (ISPs). You could choose to sign up with AOL or whomever, or use your university's or business ISP to get on the Net and go anywhere. From the beginning, the Internet designers had developed a system that would be open to innovation, but they needed that court ruling declaring open access. (It is now being rescinded, unfortunately, as we shall see.)

This interoperability is possible because all the computers agreed on rules or protocols that specify how data will be transferred. For example, the format of a Web page is specified by hypertext markup language (HTML); the order in which bits of the message are sent is specified by the hypertext transfer protocol (HTTP); there are many other protocols, such as the basic transmission control protocol and the Internet protocol, together known as TCP/IP.

The Internet protocol (IP), being minimal and general-purpose, can work with the enormous diversity of networking technologies. For example, after the IP was universally adopted, a new technology came along—local area networks (LANs), that even worked over wireless networks; the IP worked fine with it. IP is being adapted to accommodate ever newer technologies, such as asyn-

chronous transfer mode. The fact that any one of these uses, and more to come, such as telephony, can be carried on one platform makes the growth of the Internet and the World Wide Web possible.

Separate dedicated connections for each use would probably be cheaper and more reliable, just as dedicated machines in a mass-production industrial plant are cheaper and more reliable than multipurpose machines that have to be reset for each operation. But if, as we will see, you can find a way to automate everything running on the platform, and can build in a lot of redundancy to protect from inevitable failures, then the system with expandable platforms and interoperability is far more efficient in the long run. New modes of connection, replacing the slow and limited capacity of telephone lines, can be accommodated, such as fiber-optic cable, or wireless radio waves as in a LAN in your house, company, or even a town. The electric power grid, which goes to every home, may come to be utilized; and links to satellites are appearing. These connecting modes can be provided by for-profit organizations, but they have to abide by the protocols those government employees evolved decades ago, or no one would use them. However, for-profit firms are finding backdoors to this open-systems approach in order to control content and usage and increase the charges for using their "pipes" (fiber-optic cables, telephone lines, etc.).

### A LITTLE HISTORY

How did the protocols now in use come about? Initially, several commercial firms were developing and selling protocols, including IBM, Digital, and Xerox. But the government was also developing them and had a head start since the initial internet was financed by the Defense Department (with the National Science Foundation coming in later). The government protocols were free, and small groups and associations could modify them. The free protocols encouraged wide use, the wide use brought in a range of improvements and modification, and these won out over the commercial ones. Furthermore they were made to be compatible with any ven-

dor's hardware. Interoperability was to be the key. (Gillett and Kapor 1997) A privatized internet would be unlikely to encourage an IBM customer to use software or hardware from Digital; IBM would insist that its own products be used and none other, just as the Bell system had tried to do and Microsoft after that. The decision to establish rules that would allow anyone's product to work, provided that it followed the protocols that were offered free, led to what is called an open-standards source of protocols. Open standards means nonproprietary.

While the government's role in establishing the Internet was crucial, economic competition from private firms became the engine of the Internet after 1995. In 1995, the National Science Foundation allowed commercial firms to establish themselves as ISPs, to the outrage of university users who did not think it should be open to commercial use. But commercial ISPs brought the cost of access down. For example, junior colleges and other schools could then afford to participate. Soon the prices were low enough for users in their homes. Competition among ISPs has been intense, and the cost of access to the Net has fallen greatly. (Gillett and Kapor 1997)

### RELIABILITY OF THE INTERNET

How could such a vast network function with only small, sporadic failures that are more annoying than consequential. First, there are very few human beings involved in managing its operations—far fewer than the power grid network—in part because it is electronic, and electronic devices are more reliable than electro-mechanical or mechanical devices and require fewer human monitors. Human error accounts for many software failures, but these do not involve the Net. Your browser may crash, or hackers may gain access to your financial data, but the Net itself goes on. Second, there is a great deal of redundancy designed into the Internet. We will examine some of this since it offers lessons for other systems, some of which are applicable to our critical infrastructure. Third, its structure is remarkably decentralized; it is probably the

most decentralized system on earth run by humans. Decentralization is easy because of its electronic character, but mechanical and human systems can also be quite decentralized, as we shall see in the next chapter.

Why is the Internet so robust with regard to errors? Not only is it electronic, but it has built into it a great deal of redundancy. Messages are broken up into "packets," each of which carries the address they are to go to and the addresses of the other packets that make up the message. It does this through two methods that are quite different from what we normally think of as redundancy. The normal notion of redundancy in a nuclear power plant, for instance, means that there is a backup, such as a diesel generator, that can be brought into play if the normal source of power (the power from the nuclear plant itself) fails. But the Internet has additional sources, called replacement or link redundancy and mirroring. (These are complicated, and are briefly discussed in Appendix A.) Most systems have many redundancies either designed into them (the preferred form) or added on (a less reliable form for a number of reasons). But the Net's redundancy is found in a peculiar network characteristic; it is a "scale-free" system, a characteristic of some other examples of networks.

Roughly this means that nodes (computers and ISPs) can be added to the network without increasing the levels of supervisory control and thus hierarchy. It does not need an additional level of control when another thousand nodes are formed because there are an enormous number of different paths over which a message can travel. (To return to our highway analogy, if one road is blocked there are other roads you can use. In actual highways, however, there is a hierarchy; some roads are interstate highways, others country lanes. In the Internet, all roads are equally fast if not congested, so there are many more choices. [This is called net neurality.] In addition, the speed at which the message can travel is almost instantaneous.) A very high number of these paths can be disabled, but the message will still get through.

Unfortunately, physicists observe, there are a very small number of nodes that have millions of links to other nodes, and if these are

disabled, the whole network will crash. So while it is safe from random attacks, errors, and electronic failures, which accounts for its reliability, it remains vulnerable to deliberate attack on a few critical nodes. "Such error tolerance and attack vulnerability are generic properties of communication networks." (Albert, Jeong, and Barabasi 2000) However, the few critical nodes, called hubs, are currently highly defended by the voluntary groups that oversee the Net. As we shall see, this could change.

I have gone into this arcane and complex structure because it allows me to make a very important point. A scale-free system is able to increase in size, in this case, increase the number of nodes or points of connection, without changing its structure. This means that networks are not subject to the perils of concentration that have so concerned us in this book. If it were possible to reduce the vast number of ISPs (concentrating the access points of your computer to the Internet), tell you which roads you could use (limiting access to sites), and charge you more for using a faster road, the centralization of the Net would increase, and there would be nodes whose disruption by accident or intent would cause great rather than trivial damage. To see how this works requires a closer examination of the structure of the Internet and the World Wide Web.

The World Wide Web, which works through the physical network of the Internet, is also a scale-free network, in that there are a few very central Web sites that link to many other sites, while the vast majority of sites only link to a few sites. But for your computer to connect to a site on the Internet (such as www.wikipedia.org), it must know where the computer(s) that make up Wikipedia are located on the Internet. Every computer on the Internet has a four "octet" number (an IP number, such as 69.73.163.167) that routers on the physical network use to move traffic from one computer to another. So in addition to the physical hubs that connect computers on the Internet, there is a group of servers that tell computers how to find each other using the domain name system (DNS). This translates familiar names such as wikipedia.org into numbers that the routers can use to send information between these servers and your computer.

There are essentially only four levels of hierarchy in the domain name system, and thus in the Internet as a whole. At the top level are *root servers*. There are thirteen of these for the whole world (although for safety and reliability reasons they are duplicated in different physical locations). We might consider this as level one in the four-level informational hierarchy of the Internet. If a computer wants to know how to find www.wikipedia.org, it would first go to one of these servers to ask where to find a server in charge of addresses that end in ".org." Level two consists of thousands of DNS servers, which are clients of level one servers, the root servers. These clients themselves are servers for more clients below them. The thousands of servers in level two are often grouped into "server farms" having thousands of processors each. (The size of the farm, or data center, is only limited by its ability to remove the heat generated by each processor in each server. They are generally large, windowless air-conditioned buildings in major cities.) These servers may not know the address of www.wikipedia .org, but they do know which server(s) are in charge of the wiki pedia.org domain. Level three consists of servers that know the addresses of the computer(s) that make up www.wikipedia.org. These computers are located at the fourth level of the hierarchy. Once your computer knows the address of the Wikipedia servers at level four, it can contact these servers directly. This whole process is shortened by *caching:* if another computer at your ISP recently asked for the address of www.wikipedia.org, it would already know where to find it without working its way through the entire hierarchy.

There is no central body running this vast system. There are, however, some crucial organizations that "rule the root." One is something called ICANN, which establishes domain names (such as .com, .edu, or .net), Internet protocol addresses, and protocol port and parameter numbers. ICANN is a nonprofit organization in the United States, and its control is being contested by other nations that want to be able to control their own ISPs (of which more later). Others include organizations coordinating the routing of traffic on the net, setting service quality standards, protocol stan-

dards, and the unique identifiers, which are controlled by organizations such as the Internet Engineering Task Force (IETF).

A number of committees deal with proposed changes in standards, and there is an Internet Architecture Board that ratifies them. The committees, and boards like this, are open to public participation and review. The government of the Internet, and through it the Web, is light, nonproprietary, tiny in size, very expert, and largely voluntary. It is a remarkable achievement, considering the vastness of the system and its importance for modern society.

## THREATS TO THE INTERNET

The main threat to the Internet is not that it might be shut down, but that its openness, innovativeness, and nonproprietary nature could be compromised. Were it to pass into proprietary hands, the number of ISPs could decline drastically, thereby restricting its openness but also making larger, concentrated targets for thieves, terrorists, and those that are simply malevolent.

An amusing, though trivial, instance of the threat to the openness of the Internet came from U.S. Senator Orrin Hatch (R-UT), who suggested in 2003 "that people who download copyright materials from the Internet should have their computers automatically destroyed," according to the news story. He was concerned about downloading copyrighted music. But it was pointed out that he was himself using copyrighted software for which he had no license on his official Web site. (Kahney 2003) Less drastic efforts than smoking computers have been utilized to prevent pirating of music and videos, but with limited success. However, pirating of copyright materials is a small problem compared to the controversy regarding access to the "pipes," the fiber-optic cables and the high-speed phone lines that most of us use.

The problem begins with the fiber-optic cables that carry more and more of Internet traffic. They are privately owned and built by cable and other telecommunications firms. From the beginning, the courts ruled they were common carriers, that is, like highways

and ship channels that everyone has access to for a reasonable fee. In the late 1990s, firms overbuilt these pipes to such an extent that it depressed the dot-com sector, helping to bring about a serious downturn for Silicon Valley and other telecommunications sites. With so much excess capacity, the cable companies could only charge small fees for leasing space in the pipes, called *bandwidth*. But as traffic increased with more Internet users and especially with the downloading of the large files required for music and videos, the pipes became crowded and service was supposedly threatened. (Another interpretation is that once people were enticed onto the Net, larger fees could be charged to provide high revenue at little additional expense. There is still a surplus of bandwidth.)

In a crucial decision, the FCC ruled that the pipes were no longer common carriers but proprietary; they should be governed by telecommunication rules, which themselves were being relaxed. Since they are no longer common carriers, bound to let any user on that paid a reasonable fee, the owners could keep some ISPs from using them if they wished, and thus potentially control the content by denying use. The owners argued that this was desirable; they could block pornography and spam. But they could also block sites with political content they disapproved of. They can block competitors who offer music or videos or e-mail or shopping or whatever that is in competition with their own offerings of these items and services. Congress revisited the issue in 2006, with firms that are heavy users of the Internet—such as Google, Amazon, and Yahoo!—wanting continued inexpensive access, on one side. On the other side are the network operators such as Verizon and cable companies. They want to be able to block sites. They also want to be able to charge higher fees for heavy usage (not an unreasonable demand). The fear is that the carriers will block consumers' access to popular sites or degrade the delivery of Web pages whose providers don't pay extra. Google's home page, for example, "might load at a creep, while a search engine backed by the network company would zip along," according to a *Business Week* story in 2005. It stated:

[R]ecent court and regulatory rulings have given the carriers more room to discriminate. In June, the U.S. Supreme Court ruled that cable broadband services were almost free of regulation. Two months later, the FCC granted the same liberty to the Bells' broadband services. The FCC made two newly merged megaphone companies—created from AT&T and SBC and Verizon and MCI—vow to keep their Internet lines open to all for the next two years. But FCC Chairman Kevin J. Martin favors a light regulatory touch until he sees widespread abuse by the networks. (Yang 2005)

But the issue of volume pricing is also not settled. It is estimated that file-sharers swapping music and movies accounts for 60 percent of North American residential broadband use. As the Net predictably slows with congestion, rather than add more fiber-optic cables, the carriers may prefer to charge heavy users more, opening the door to slowing down the pages of competitors. A Korean phone giant found that 5 percent of its users accounted for half of its traffic. (Yang 2005) Just as adding more transmission lines or upgrading them has not been in the interests of the newly consolidated electric power industry, it may not be in the interests of broadband owners to increase supply when it can get higher revenues without the capital costs.

"The end of the Net as we know it" has long been a fear. A variety of nonprofit, commercial, and government organizations governs the Internet, but its astounding growth has made it a prime commercial target. In 2003, VeriSign, the company that controls two of the thirteen root DNS servers, .com and .net (they connect Web browsers with sites fourteen billion times a day!), argued that the infrastructure should be commercialized by getting rid of the nonprofits. The head of VeriSign told reporters: "It's time for the internet infrastructure to go commercial. . . . it's time to pull the root servers away from volunteers who run them out of a university or lab. . . . That's going to be an unpopular decision." (Murphy 2003) VeriSign did not succeed, but the threat remains.

In 2003, Michael J. Copps, a Federal Communications Commissioner, expressed his fears about the centralization taking place in telecommunications:

> Once upon a time, cable was going to save us from too much network control of the broadcast media. Today 90 percent of the Top 50 cable channels are controlled by the same corporations that own the TV networks and the huge cable systems. Then we were told not to worry because the Internet would be the ultimate protection. We looked at the top 20 news sites on the Internet. Guess who controls most of them? The same big companies that provide us with our TV and newspaper news. Some protection. (Copps 2003)

## CONCLUSIONS

The reliability of the devices that access the Internet is in question, as we saw earlier, and much damage could be done to the nation because of their lack of security. Military establishments have been broken into and their passwords taken, making the military vulnerable to foreign countries and terrorist groups; the control systems of electric power plants and transmission facilities have been penetrated (though as yet without any damage that has been publicly admitted); nuclear plants have been disabled for hours; the FBI's computer system was temporarily rendered almost inoperable; and the amount of criminal activity involving money and industrial secrets is said to be extensive. Most of the vital things in our nation's critical infrastructure now depend on Internet security, down to the automated hospital systems for matching prescriptions with patients and delivering them. Anything that skimps on reliability or security in this vast system endangers us, and there are two things that allow skimping: monopoly power by a for-profit enterprise, and concentration of the critical nodes of the Net. The first occurs because reliability and security cost money for the monopolist, and that reduces profits. The second, concentrated

nodes, occurs because concentration of ISPs translates into market power, and the consequence of this is a larger and more vulnerable target for terrorists to attack, or for industrial accidents and errors to disrupt.

The main problem identified in this chapter is the hegemony of Microsoft operating systems whose complexity and proprietary safeguards have created products that lack reliability and security. Of course this is an industry problem, and if a firm like Google were to get 95 percent of the market share and did not have to pay attention to reliability or security because of it, we would be in the same position. But since the industry is dominated by a single firm, Microsoft, it makes little sense to criticize the industry and not the firm that is responsible for the shape of the industry.

But the most basic criticism is of a society that allows such market concentration to exist. We need broad antitrust regulations that say it is not in the public interest to have a single firm dominating a market because it can lead to monopoly profits, and if it is critical to our infrastructure, to increased vulnerability. Competition in telecommunications will lead to more innovations and to systems that compete on the basis of reliability and security. Our present judicial ideology is unable to address this issue. After a court ruled that Microsoft had clearly engaged in anticompetitive behavior, Microsoft appealed the decision. An appeals court accepted the ruling of anticompetitive behavior but levied penalties that were trivial, and most important, declined to break up Microsoft in such a way as to encourage competitive operating systems. We can only hope that the development of competing virtual operating systems that may be accessed from the Net might make security a prominent competitive feature. Or that more states and even the federal government might do what Massachusetts has done (to save money) by requiring open sourcing of its operating systems and software.

A correspondent at Microsoft puts it well in a personal communication: "[A]s a society we have created a world where companies routinely under-estimate, and under-invest in, risks, and many systemic risks are externalized in such a way that nobody deals with

them. It is a society where technical people are unintentionally complicit in this by providing overly-optimistic analyses, blaming the user, and avoiding taking responsibility for dealing with the real limitations of all systems today."

The reliability of the Internet itself is not in question. It is a model of a highly reliable, secure, efficient, decentralized system of vast size. It is the security of the devices that use the Net that is the problem at present.

However, the governance of the Internet raises two concerns. First, it is the freedom of access and content that has made this such an extraordinary system in a world where freedoms appear to be diminishing. If AOL or Comcast can restrict your access to sites on the Internet that they have a commercial interest in, or an ideological interest in, the present freedom that we enjoy will be constrained.

More important for the analysis of this book, the reduction in the number of ISPs that are available, and the number of sites that can be accessed by those ISPs, will constitute a centralization of what is at present a remarkably decentralized system. The security issue enters here. The consolidation of servers will also make them larger targets for terrorist and hackers and more vulnerable to human and electronic failures, and perhaps even to weather or other natural disasters, though that seems remote. Once again we see that concentration threatens the security of our critical infrastructure. The market for security has failed or not been enough of an urgent concern. Government must play a role; as the largest purchaser of software it could insist that reliability be a condition of purchase. The courts could legitimate suits around security interests, thereby bringing insurance into play and supplying another incentive for secure products.

The few movements in this direction are only concerned with consumer security, not the security of our critical infrastructure. Still, it is encouraging that the Federal Trade Commission brought unfair trading practices actions against Microsoft and Eli Lilly, claiming they had misled consumers with their promises of the security and privacy of customer information. Also California has

required businesses to disclose computer security breaches if they allow unauthorized access to personal information. (Vadis 2004, 109–10) Perhaps, as with the initial consumer-oriented efforts to get the attention of the chemical industry, these efforts, now concerned with competitiveness, freedom of choice, and privacy, will expand into areas where there is the potential for large disasters.

# PART 4

## What Is to Be Done?

----------------------

# 9 The Enduring Sources of Failure

## Organizational, Executive, and Regulatory

-------------------------------------------

THIS BOOK HAS BEEN ABOUT the inevitable inadequacy of our efforts to protect us from major disasters. It locates the inevitable inadequacy in the limitations of formal organizations. We cannot expect them to do an adequate job in protecting us from mounting natural, industrial, and terrorist disasters. It locates the avoidable inadequacy of our efforts in our failure to reduce the size of the targets, and thus minimize the extent of harm these disasters can do. But why are organizations so inadequate? First, there are the inevitable human failings of cognition, motivation, organizational designs, and so on, and the unpredictable and often hostile environment the organizations have to work in. This is *organizational failure* per se, as when workers and management fail to do their jobs for whatever reasons, or the jobs make demands beyond their resources. We saw this in Katrina when first responders failed to take obvious steps to aid people that they could have taken, and when the size of the storm overwhelmed responders. The failure of power suppliers and even the Department of Defense to isolate their sensitive operations from the public network, thus inviting hackers to do damage, is another. Organizational failures are one explanation for the failures of FBI headquarters to follow up on warnings by field officers of the possibility of terrorists using jet-

liners as weapons, the firing of a translator that warned of mis-
translations by a coworker who was tied to terrorist organizations
and the subsequent promotion of the translator's superiors, or the
failure of the Clinton administration to correct misunderstandings
about the firewalls between the CIA and the FBI. More commonplace is the mismanagement of contracts by government agencies;
it can affect our ability to respond to disasters. For example, a
multibillion-dollar contract to set up a computer system for the FBI
was poorly planned, frequently changed, and not given proper
oversight, resulting in extensive waste. (McGroddy and Lin 2004)
Lack of agency oversight enabled the Unisys company to overbill
and supply faulty equipment to the Transportation Security Administration for a communications system, where $1 billion "went
down the drain." As with FEMA in the Reagan administration,
TSA ended up with ancient RadioShack phones and little Internet
capability. (Lipton 2006a)

Organizations are hard to run; people don't always do what they
are supposed to do. They also reflect diverse, conflicting external
interests and diverse, conflicting internal interests. Information
and knowledge is always insufficient, and the environment is often
hostile and always somewhat unpredictable. Thus, there is the
ever-present problem of prosaic, mundane organizational failure.
(Clarke and Perrow 1996) We certainly should work hard at improving our organizations, and there is a large industry of consultants, college and university programs, workshops, and books at-
tempting to do this. But just as I have argued that our response to
the three sources of disasters needs to be something more than at-
tending to preparation, response, and mitigation, we need to do
more than improving the functioning of our existing organizations; our efforts there can only result in minimal improvements.

Another internal source of organizational failures is what I
have called *executive failures,* where top executives make deliberate, knowing choices that do harm to the organization and/or its
customers and environment, as evidenced by the Millstone nuclear
power plant case. It is important to contrast this with mundane
organizational failures, which is the more favored explanation.

When organizations with disaster potentials have accidents or near accidents and many lives are threatened or taken, the response has been to blame the operator, or perhaps the organization and particularly its lack of a "safety culture." Changing the culture of an organization to correct its ways has become a mantra in the industry that sells organizational-change techniques, and also in the academic field of organizational behavior, where concerns with culture, rather than structure and power, dominate. But organizational cultures are very resistant to change, as the heads of NASA after the *Challenger* and then the *Columbia* disasters can attest. The independent investigations of both these disasters cited a "failed safety culture" as the main problem. Identifying this after the first disaster produced no corrections that could have prevented the second one. The same diagnosis was given in the Millstone and the Davis-Besse nuclear power cases, which we reviewed. Why such failures? I think this is because the executives at the top have other interests in mind, and their interests would be threatened by an effective safety culture. This points toward executive failures.

Classifying something as an executive failure rather than a mistaken executive strategy or a poorly performing executive, or even a failure by management or workers, is controversial. Observers can disagree, and since it is easy to have a mistaken strategy or a poorly performing organization, it takes a great deal of evidence to make the case for executive failure. In fact, only when executives are charged in courts are we likely to get the evidence needed for executive failure, as in the Enron and Millstone nuclear power plant cases, but one can easily imagine it plays a large role in the Davis-Besse case. I believe that the administration of George W. Bush was responsible for executive failures in the decision of the president and a handful of top officials to ignore the well-documented threat of non-state-sponsored terrorism and focus instead on threats of state-sponsored terrorism, planning a possible invasion of Iraq (prior to 9/11), mounting a defense against nuclear weapons in case of an attack, and implementing massive tax cuts. These efforts left us more unprepared for the 9/11 attack than we were under the

previous administration. But it is only my judgment that this was an executive failure rather than a mistaken strategy, and it cannot be proved. The Clinton administration might have done just as poorly. But bluntly dismissing warnings and cutting the budget requests of the agencies that were alarmed and wanted more resources sounds like more than a "mistake" to me, and more like an executive failure.

The failure is to ignore warnings of disaster that their own experienced personnel sound. Millstone executives did not *want* to nearly cause a meltdown; the Bush administration did not *want* to bring about a frightening attack by terrorists. To say they intended these disasters is to support far-out conspiracy theorists who have argued that President Roosevelt wanted the attack on Pearl Harbor in order to bring the country into World War II, and that President Bush ignored the warnings of a terrorist plot to use airplanes as weapons in order to justify his planned attack on Iraq. These charges clearly lack credibility.

If executive failures create opportunities for organizations to do harm—and we should expect executive failures because nothing is perfect—the remedy is to replace the failing executives. This is possible (if difficult) in the case of government organizations where either the executive in question is elected or the elected official is responsible for appointing executives below him or her. But in the case of corporations it is extremely difficult. The board of the corporation appoints the executive, but they effectively, if not literally, appoint themselves or people very much like them as board members. (Khurana 2002) Management and workers in the organization do not have a vote in the election of corporate leaders, nor do the communities affected by the organization. Pension funds and other large holders of stock are increasingly challenging some corporate decisions, but this is a long way from workers or communities going to the polls to choose corporate executives. My point is that for-profit organizations such as corporations are not designed to be democratic, so regulations, and then lawsuits when regulations are violated, will have to be the primary defense against executive failures. Since failures are inevitable, this is yet another rea-

son to have many smaller organizations than one or few very big one; the consequence of any one failure will be less. There are more executives who may fail, of course, but there will be less collateral damage from a failure in a small chemical plant than in one with multiple, tightly coupled vulnerabilities. But, for something like a nuclear power plants, we have to rely on regulating it rather than downsizing it (though deconcentrating the industry would reduce the pressure on the regulators from the industry's friends in Congress).

Stepping outside of the organization itself, we come to a third source of organizational failure, that of *regulation*. Every chapter on disasters in this book has ended with a call for better regulation and re-regulation, since we need both new regulations in the face of new technologies and threats and the restoration of past regulations that had disappeared or been weakened since the 1960s and 1970s. The regulatory potential for avoiding disasters and reducing their consequences is obvious. Standards exist, or should exist, concerning the height and strength of levees, building standards, and the reliability and security of key parts of our infrastructure, such as the power grid and the Internet as well as our transportation system and ports. There may even be some existing regulations regarding the quantities of hazardous materials that are stored that could be strengthened. Regulation frequently fails because of political processes, but we have done better in the past and could do better in the future. However, since this enters the thicket of politics, I will postpone a discussion of regulation for later. As difficult as it will be, improved regulation has a greater chance of success than either organizational reforms or the reduction of executive failures.

Given the limited success we can expect from organizational, executive, and regulatory reform, we should attend to reducing the damage that organizations can do by reducing their size. Smaller organizations have a smaller potential for harm, just as smaller concentrations of populations in areas vulnerable to natural, industrial, and terrorists disasters present smaller targets. Smallness is itself is a virtue if we are dealing with concentrations of hazardous materials or concentrations of populations in risky areas,

two of the sources of danger that we have examined. But a third source of danger involves parts of our critical infrastructure. Here, concentrations of political and economic power are central facts and difficult to change. Power involves dependencies; safety and security in the infrastructure involves interdependencies.

## DEPENDENCY AND INTERDEPENDENCY IN OUR INDUSTRIAL STRUCTURE

It is a commonplace that we live in a highly interconnected society, but in the case of disasters the connections are largely ones of *de*pendency rather than *inter*dependency. For our vital Internet to be secure, we should not be dependent on one operating system, Windows, for 90 to 95 percent of our computers. Mergers in the chemical industry and the transportation industry have also increased our dependency on fewer organizations. Concentration in the electric power industry has made us dependent on fewer large generating sources (even though we have added many tiny ones) and has encouraged business practices that make us more dependent on long-distance "wheeling" of electricity.

There are other dependencies. The largest power generator in Missouri is the New Madrid Power Plant. It burns 16,000 tons of coal every day and requires railway shipments every 1.4 days. But all of this coal comes from Wyoming and all of it crosses the High Triple Bridge over Antelope Creek, creating a "choke point," or single point of failure for the whole system. (The coal comes from the Powder River Basin in Wyoming, a major source of coal for the nation, and 56 percent of which travels over the High Triple Bridge.) The bridge could be disabled by flooding, by a railway accident, and certainly by terrorists. Another obvious dependency is the one everyone has on the electric power grid. Power generated in the Northwest travels to California over transmission lines that have a critical link, or choke point, on the Oregon-California border, and when that failed because of overheating from overloading

in August 1996, the West had a massive eleven-state blackout.[1] In addition to backup generators or wind or solar power facilities to reduce our dependence on the grid, we need to avoid critical links that can be the single point of failure in our transmission system.

Our society would also be more secure if we were not so dependent on single modes of communication. When telephone lines went out during the World Trade Center disaster, it was fortunate that some people in the affected area had cell phones, BlackBerry devices, pagers, and other specialized forms of communication independent of those lines. It is common for multiple suppliers to emerge during disasters, such as private boats and commandeered barges during the Katrina disaster, and all sorts of means of water transportation as an estimated 500,000 people at the tip of Manhattan when the World Trade Center collapsed performed the largest water evacuation in our history (all occurring spontaneous and voluntary without any central direction). (Kendra, Wachendorf, and Quarantelli 2003) In industry, flexible multipurpose machines and multifirm suppliers and customers reduce dependencies. More important, these involve more multinode, complex networks, which are partially self-regulating, as we saw in the power grid and Internet self-organizing features. More important still for the argument on vulnerabilities, this also results in smaller concentrations of hazardous materials and lower economic and political power for any one organization.

Most of the examples above are those of physical dependencies. Another prominent form of dependency I call *spatial dependency*, drawing on the work of three engineers. (Rinaldi, Peerenboom, and Kelly 2001) (For a more analytical discussion of the two types of dependencies that predominate in our society but ought to be avoided, physical and spatial, and two types of interdependencies that are to be favored, reciprocal and commonality, see Perrow 2006.) An example occurs when the failure of a electric power

---

[1]The examples come from a PowerPoint presentation by Ted Lewis. See other illustrations in his remarkable book *Critical Infrastructure Protection in Homeland Security: Defending a Networked Nation* (2006).

source not only shuts down the nuclear reactor but shuts down the cooling system needed to keep the spent fuel rods from fissioning. The two systems—power generation and spent-fuel cooling—need not be linked but are, for minor economic reasons. There are clear economic efficiencies associated with most spatial dependencies. But it would cost only a little bit more to move the spent storage pool to a distant site that does not depend on the nuclear power plant for power (price should not be the only measure of utility). A power failure could crash both the plant and the pool cooling system, and emergency generators frequently fail and soon run out of fuel. (A terrorist attack could also bring down both tightly coupled systems, so keeping the spent rods cool is considered by some to be the major vulnerability of nuclear power plants.) (Alverez 2002) Where catastrophic failures might be experienced, as in the nuclear power case, the risks seem very high compared with the economic benefits.

Many spatial dependencies are so thoroughly built into our constructed environment that little can be done about them. A chemical plant explosion that also disabled the water supply of the town that was needed to fight the fire the explosion caused is an example we have encountered. The collapse of 7 World Trade Center and its Verizon facilities illustrates spatial dependencies. There were 34,000 customers who lost telephone and Internet services, many of whom thought they had redundancies because they did business with multiple carriers. But most of their lines ran through the Verizon central switching office, which collapsed when one of the towers fell on it. Of course, using multiple switching offices would be more expensive, and how often do you expect such a disaster? So the tight coupling of spatial dependency has been largely restored in the new Verizon facility. (Blair 2002; Council 2001)

Concentrated industries have concentrated facilities, by and large, with many spatial dependencies, and some are likely to have hazardous materials. Concentrations not only increase the destructive power of hazardous materials, as with chemical plants, but can increase the spread of toxins, whether the toxin comes from nature or from terrorists. A National Academy of Sciences re-

port says that while the deadly bird flu virus, H5N1, began in wild birds, it developed its power to spread because of the cramped conditions of Asian factory farms. It is factory farming and the international poultry trade that are largely responsible for the spread of bird flu. (Lean 2006) Concentrated factory production methods are also held to favor the spread of foot-and-mouth disease and mad cow disease in feedlots in the United States. (Kosal and Anderson 2004)

Terrorists are unlikely to choose poultry farms or feedlots as a target, but they could very easily choose another point of concentration, the massive silos that hold raw milk. Lawrence Wein and Yifan Liu (2005) note that the release of botulinum toxin in cold drinks, including milk, is one of the three most fearsome bioterror attacks (other than genetic engineering of toxins). In their model, a botulinum toxin is deliberately released in a holding tank at a dairy farm, or a tanker truck transporting the milk to the processing plant, or directly into a raw-milk silo at the processing plant. It does not matter which, the toxin eventually will be well mixed in the silo. In the typical plant, the silo is drained and cleaned only once every seventy-two hours. A silo can hold 50,000 gallons and several in California hold up to 200,000 gallons, a splendid example of concentration. In their model, if only ten grams of a toxin is put into one of these sources, the mean number of people who will consume contaminated milk is 568,000. It would take from three to six days for the casualties to appear. Children are especially likely to die because they consume more milk and are more vulnerable. We would not necessarily have a half-million deaths from just one contamination, and botulism is not necessarily fatal even for children, but even simply disabling a half-million people would have tremendous ramifications. The surge capacity of our hospitals would quickly be overtaken, and the psychological impact of an attack on milk cannot be overestimated. Other possible targets for such an attack are fruit and vegetable juices and canned foods, such as processed tomato products.

The milk industry is aware of its vulnerabilities, but all prevention methods are, as we might expect, voluntary. The methods

involve locking tanks and trucks, having guards at each stage, inspectors, and improved pasteurization. (Wein and Liu 2005) Since the appearance of their article (first rejected by *Science* magazine at the request of the government), Wein, in a personal communication, notes that the industry has taken further steps. But the simplest method is to reduce the size of the targets; terrorists would not find them as attractive, and industrial accidents would be limited to harming fewer people. Are the economic advantages of concentrated milk processing so overwhelming that we should not consider limiting silo size?

Or consider hazardous materials on railroad cars on heavily used transportation corridors. Rerouting ninety ton railroad tank cars away from urban centers seems obvious, but it was done only temporarily for Baltimore and Washington DC. The federal government could require that fewer hazmats be carried in large quantities on any single freight haul—to spread them out, as well as require safer freight cars and tracks. No new laws are required; the government already has the authority to require stronger containers to replace the aging ones, to require that they withstand grenades and other terrorist weapons, to reroute shipments, and to require simple tracking devices that would allow local police and firemen to know what they have to contend with. When New Jersey tried to regulate in this area the Supreme Court upheld the railroads' assertion that only the federal government can regulate rail traffic, a catch-22. (Kocieniewski 2006) (Why doesn't the federal government do more safety regulation?)

The ideal industrial structure that reduces dependencies is one that has multiple producers that draw on multiple suppliers and sell to multiple customers. If anyone of these fails, there are alternative sources of producers, suppliers, and customers. Multiplicity promotes competition within each of the three factors. If customers demand reliability and security in their operating systems, multiple producers are more likely to emphasize reliability and security than, say, prices or features. With a single dominant operating system, this choice may not be available.

An ideal industrial structure also should promote interdepen-

dencies as well as reduce dependencies. Interdependencies involve reciprocal coordination where, for example, the supplier can recommend to the producer changes in his product that would make the supplier's task easier (and price lower) and even improve the product. The producer should be able to do the same with the customer, getting the customer to change her requirements to facilitate the producer's tasks. (The best salesperson is the one that is able to alter the behavior of the customer to the benefit of both the supplier and the customer.) The incentive to engage in interdependent, reciprocal, and even power-sharing relations is most likely to appear where there are multiple customers, producers, and suppliers. Monopolies do not encourage this kind of behavior. Having many, rather than a few, organizations reduces the size of the targets, creates redundancies that can better handle extreme events, and also reduces the dangers of choke points.

The more ISPs in the Internet, the more redundant pathways will be present and the less defended any one of them will need to be. Choke points threaten many parts of our critical infrastructure. In addition to the transportation of coal from Wyoming and much of California's energy from the Northwest, there are choke points on oil and gas pipelines (Lewis 2006, chap. 10) and on evacuation routes from hurricane-blasted coasts. The answer to this problem is having multiple connections so that fewer nodes are critical. Multiple connections are more likely to be produced where there are multiple producers and customers. Choke points are the result of concentrations, and many of these are large firms with monopoly or near-monopoly power.

Barry Lynn (2005) applies this kind of analysis to the world industrial system, rather than just the critical infrastructure in the United States. By eliminating the "bulkheads" that traditionally existed between companies and even between nations, which protected them from shocks and distributed risk, globalization has increased single-point dependencies in the search for efficiency and low prices. A war on the Korean peninsula would remove half of the global production of vital D-ram chips, 65 percent of Nand flash chips, and much more. The electronics industry would be dis-

rupted along with the many industries that depend on it. An earthquake in Taiwan in September 1999 caused a weeklong break in the island's electrical and transportation systems. Production ceased and inventories of key computer goods ran out in the United States, closing thousands of assembly lines and tumbling the stocks of major electronic firms such as Dell, Hewlett-Packard, and Apple. The world output of electronics was cut 7 percent below predictions in just the next month, and at Christmas time in the United States there was a shortage of computers and electronic toys. (Lynn 2005) In an article he shows how Wal-Mart in particular has promoted large, sole-source production in developing nations, forced its U.S. suppliers to outsource, and fought all attempts to increase port security. (Lynn 2006) In his book, Lynn writes:

> Our corporations have built a global production system that is so complex, and geared so tightly and leveraged so finely, that a breakdown anywhere increasingly means a breakdown everywhere, much in the way that a small perturbation in the electricity grid in Ohio tripped the great North American blackout of August 2003. (Lynn 2005, 3)

I couldn't have put it better!

## NETWORKS OF SMALL FIRMS

There is an alternative model for relations between organizations, first theorized about fifty years ago after economic demographers in Italy discovered a strange phenomenon. Northern Italy had a very large number of small firms, and while this should have been associated with low economic development it was actually associated with high economic development. Many of the firms were in high-tech industries and leaders in their field, but they generally had fewer than twenty employees. They were in increasingly prosperous areas, with low rates of unemployment and high rates of access to child care and higher education for females. Since its first theorization, there has been a burgeoning literature on the effi-

ciency, resiliency, reliability, innovativeness, and positive social outcomes of networks of small firms in a variety of countries. Northern Europe, Japan, Taiwan, and even parts of the United States (for example, Silicon Valley and centers of biotech excellence) have seen their appearance.[2]

Networks of small firms include the core of producers, some distributors (which may also engage in production), and a range of suppliers, most of them small themselves. With multiple producers, distributors, and suppliers, there is a great deal of inexpensive redundancy in the system. Dependencies are low because there are multiple sources of supplies, producers, customers, and distributors. The failure of one firm (whether for business reasons or one of our disasters) does not disrupt the interdependencies, since other firms in the network can easily expand to absorb the business. Firms have multiple suppliers and multiple customers, reducing dependencies. With few investments in single-purpose equipment or layout, changes in customer requirements are easily met. With many links to the environment, new innovations and changing demands are readily sensed. Personnel move between firms easily (even without changing their car pool), carrying with them the innovations that are now "in the air" and keeping the whole network up-to-date. Wealth is decentralized, since it is spread over many units, and thus the economic power of individuals or single units is kept in check while the power of the network is enhanced. (Appendix B discusses some of the issues associated with networks of small firms that are raised by critics, such as transaction costs, the range of products, and threats from large organizations.)

But can this be a model for chemical plants, nuclear power plants, the electric utility industry, and all the other parts of our

[2]There is a vast literature on networks of small firms. The classic book that gave the notion, discovered by Italian demographers, its first prominence, is by Michael Piore and Charles Sabel. (Piore and Sable 1984) The concept is developed by Mark Lazerson in Lazerson 1988 and Lazerson and Lorenzoni 1999, and is illustrated in a study of the garment industry by Uzzi (1996a; 1996b). For a synthetic review, see Perrow 1992. For a more recent extension, see Amin 2000. For an application to nineteenth-century America, where networks of small firms were much in evidence, see Perrow 2002.

critical infrastructure that are vulnerable? Clearly not, in the sense that the chemical industry or nuclear power plants cannot be turned into networks of small firms. But partial steps can be taken in this direction. The nuclear plant's spent–fuel rod pool can be decoupled and moved, the giant chemical plant installation could be replaced with several smaller ones. Though it does not figure in our critical infrastructure, our giant steel mills were successfully replaced in many cases by mini mills that have some of the characteristics of networks of small firms. Chemical plants might be similarly reconfigured. If the externalities associated with their operation were taken into account, giant feedlots and poultry complexes would prove to be more inefficient than small-scale production. Parts of Silicon Valley are already made up of networks of small firms; some of the large firms in the valley function much as networks of small firms. (Saxenian 1996) And as I have argued, operating systems for the Internet would be more reliable and secure if they were not dominated by one large firm. Unfortunately, examples of networks of small firms would also have to include terrorist networks, and it is to these that we will now turn.

## TERRORIST ORGANIZATIONS

### Networks of Small Cells

Unfortunately many of the resiliencies, economies, and decentralization that characterize networks of small firms are found in the radical Islamic terrorist organizations.[3] There are estimated to be thousands of cells throughout the world at least loosely tied to the Al Qaeda network. They are very reliable. Rounding up one cell does not affect the other cells; new cells are easily established; the loosely organized Al Qaeda network has survived at least three decades of dedicated international efforts to eradicate it. There are

---

[3]Portions of this section are adapted from Perrow 2005.

occasional defections and a few penetrations, but the most serious challenge to the Al Qaeda network was a lack of a secure territory for training, once the Taliban was defeated in Afghanistan. But the invasion of Iraq has provided an even better training ground.

Terrorist networks can live largely off the land, can remain dormant for years with no maintenance costs and few costs from unused invested capital, and individual cells are expendable. There are multiple ties between cells, providing redundancy, and taking out any one cell does not endanger the network, just as in the Internet and the individual generating plants in the power grid. Reciprocity is presumed to be very high. Devotion to a common religion ensures a common culture (rather like the IPs in the Internet), even if the primary motivation is not religious. Like the power grid, the Internet, and small-firm networks, there are at most four levels of hierarchy in this vast network of terrorist cells. (For a more systematic comparison of these four forms see Perrow 2006. For a recent book with good coverage of terrorism, and another of its network form, see Pape 2005; Sageman 2004.)

As impressive as these organizations are, it is important not to exaggerate their effectiveness. For example, these are organizations and like all organizations, terrorist organizations also make mistakes. While one must be impressed with the 9/11 terrorists' extensive planning, many earlier terrorist plots were bungled. A fire in a terrorist's apartment led to the discovery of the 1995 plot to blow up several passenger jets over the Pacific; an ill-trained driver led to the discovery of the plot to blow up the Los Angeles airport in December 1999. Terrorists responsible for the bombing of the World Trade Center in 1993 were arrested because one of them insisted on getting back a $400 deposit on the van used in the bombing, which he claimed had been stolen. On several occasions the 9/11 plot came very close to being uncovered because of sloppy preparations or trivial infractions that caught the attention of authorities. They were even indiscreet in their places of employment. (These examples come from the report of the 9/11 Commission. [Staff 2004]; see chapter 4, above.) An FBI agent tried to warn his superiors of a suspected plot to fly an airplane into the

Twin Towers for two months prior to 9/11 but was rebuffed. Simple failures by the terrorists led to his suspicions. (Sniffen 2006)

One failure of the 9/11 terrorists was surprising. On United Airlines flight 93, on which the passengers fought the hijackers, the hijackers knew passengers were making cell phone calls but they did not stop them. Either it did not occur to them that the passengers might be aware the planes were being used as weapons, and foil their plan, or they could not conceive of the passengers attacking them. This failure saved either the White House or the Capitol. (Staff 2004, 12) The passengers' response, emergent and heroic where so much else was bureaucratic and bungled, was the *only* act that day that reduced the damage.

Chapter 2 of the 9/11 Commission report gives a compact history of Al Qaeda, and in organizational theory terms it should not leave us trembling because we think this group is a unique, invulnerable weapon; it is after all only an organization with all the problems of organizations, especially before its present network form was fully realized. Ideology is the key component, but not all of the core swore fealty to bin Laden; they had problems with corrupt staff, and there were signs of all-too-human disaffection. (For a striking analysis of the familiar organizational problems that affect even terrorists groups see Shapiro [forthcoming].) Al Qaeda had its ups and downs, like many organizations, and changed its structure when needed. In Sudan, prior to its expulsion and move to Afghanistan in 1996, the organization had a familiar holding group or franchise structure; it concentrated on providing funds, training, and weapons for actions carried out by members of allied groups. It changed tactics in the more friendly environment of Afghanistan, then controlled by the Taliban. Here it could centralize, and it did—with a clear leader, the equivalent of an executive committee, and specialized units for finance, training, operations, and media relations. The report says the effective, well-coordinated attacks on the U.S. Embassy in East Africa in the summer of 1998 were different from the previous ones. They had been planned, directed, and executed under the direct supervision of bin Laden and his chief aides. (67)

Subsequent disruption of the organization by the U.S. invasion of Afghanistan and the collapse of the Taliban, and the loss of perhaps a third of the leaders that we knew about, led to the vastly decentralized operation assumed to exist today. Major targets were apparently still chosen by bin Laden, but his staff and key operatives sometimes forced him to change his mind and plans, and there were divisions within the leadership as to targets; for example, some wanted to emphasize Jewish targets, others, imperialism in general. (150) (One of the few mentions of the Palestine-Israel conflict, which some believe to be a driving force for many of the terrorists, occurs in this connection in the report; unfortunately the commission was not about to visit that crucial aspect of our foreign policy, and there are few references to it. For a criticism of this aspect of the commission report, see Mednicoff 2005.) In short, the organization ran a loose franchise initially; then it centralized when the conditions were favorable and they needed an intensive and large-scale training program; and when that was more difficult, it increased its dispatch of trained agents into the world to organize and proselytize, creating the detached cells that our quite attached, bureaucratic counterterrorist organizations find so evasive and welcoming alliances with several other terrorist groups. It is presumed to have become ever more decentralized in the last few years.

There are resemblances to some Western organizations, such as direct sellers and marketing organizations, evangelical groups, and insurgent political parties. We could include criminal groups, but the important difference from these is the need for secrecy in terrorist and criminal groups. Cells were informed about plans and each others' role strictly on a "need to know" basis. Here many dots were *not* to be connected. Apparently, only the pilots among the 9/11 hijackers knew of their mission until they boarded the four airplanes. Our intelligence services and counterterrorism groups within them may have to match this adaptive, decentralized, economical, and efficient structure to meet the challenge (while operating within the law, which terrorist groups do not). Given the unwieldy bureaucratic structures in which our organizations work,

so well denounced by the 9/11 Commission, and the necessary legal restrictions, this will be very difficult. Unfortunately, the commission does not recommend decentralization to match the enemy's; as we have seen, it prescribes centralization of all intelligence functions, rather than giving our presently dispersed units more leeway, diversity, and voice.

### Setting Up the 9/11 Attack

The 9/11 Commission's account of organizing for the attacks is compelling. Those of us who wonder why Al Qaeda has not struck again should contemplate the report's account of the difficulties of the operation. The terrorists were hardly automatons. One, slated for flight training in Malaysia, did not find a school he liked, so he worked on other terrorist schemes. Headquarters found out, recalled him to Pakistan, and then sent him to the United States for flight training in Oklahoma. (225) The operation ran into many other difficulties. One hijacker could never learn English and was not successful in enrolling in a flight school. The flight instructor said he evidenced no interest in takeoffs and landings, only control of the aircraft in flight, and only in Boeing jets. (Airbus jets, the terrorists believed, have autopilot features that do not allow them to be crashed.) (245) He went back to Yemen. Khalid Sheikh Mohammed, who organized the 9/11 attack, wanted to drop him from the plan, but bin Laden said no. (222) (He did not become one of the hijackers.) There were problems with girl friends, and tensions between Mohammed Atta, a key pilot, and another pilot that almost delayed the attack because without the second pilot, they would have to use only three planes. (248)

Flight training was not easy. Hani Hanjour, who piloted one plane, was rejected at a school in Saudi Arabia, went to the United States and started at two flight schools there, stopped, and went to a third. In three months there, he earned a private license, and several more months gave him a commercial license. Then he was rejected at a civil aviation school in Jeddah, Saudi Arabia (higher

standards?), and went to Afghanistan for terrorist training. He then traveled about, ending up in Arizona a second time, struggled through a program on a Boeing 737 simulator, with his instructors saying he should give up, but completed it in March 2001, well in time to pilot American Airlines flight 77 into the Pentagon. (226–27) One can only be impressed with the determination of our enemy, but as I keep repeating, all formal organizations are in part failures.

Why have they not struck the United States again (as of this writing, in August 2006)? We should not think of terrorist groups as smooth-running machines that are quite unlike our bungling bureaucracies; they have many of the ups and downs and problems that ours do. They are often foiled, or they fail of their own mistakes; there is dissent in the leadership (Staff 2004, 250–52), and it takes years of preparation to bring off a successful major attack. But there is precious little comfort in that, and it does not explain why smaller attacks requiring less planning and coordination, which are frequent abroad, have not occurred in the United States. Unfortunately, the report does not discuss this issue, though I shortly will.

### Putting the Terrorist Threat into Perspective

Throughout this book, I have implicitly given equal rank to the threat of natural, industrial, and terrorist disasters. However, natural disasters have been more consequential, predictable, and frequent than the other two. I think the threat of industrial and technological disasters is increasing because of increasing concentration, but we have not had any disasters that have come close to 9/11. (We have come close however, as discussed in the chapters on chemical plants and nuclear power plants.) The threat of a terrorist attack from abroad is serious and we should do whatever we can to avoid and defend against one, but because this is a "strategic adversary," there is less that we can do in the way of defense than with natural and industrial disasters. A strategic adversary

can evade our defenses in a way that no flood or industrial error can. Thus, with terrorism, reducing the size of the likely targets is all the more important.

The threat of radical Islamic terrorism is only two decades old, so it is hard to compare it with the threat of domestic terrorism. But domestic terrorism is a constant in our society, and in addition to blowing up buildings, as they did in Oklahoma City, they have attacked targets of more concern to this book, such as nuclear power plants, the power grid, huge propane tanks in heavily pop-ulated areas, and the Internet. If international terrorists were to evaporate, we would still have to consider reducing such targets.

Reducing the size of our vulnerable targets is truly an all-hazards approach, in contrast to the all-hazards approach, concerned with prevention, mitigation, and response, of our Department of Home-land Security. Defending their emphasis on the terrorist threat, officials in the department argued that defensive measures against terrorism make us safer from other disaster sources. Some of the programs do apply to all hazards; more emergency-response ser-vices such as health personnel and hospitals will help in a hurri-cane, epidemic, or a chemical plant explosion. So will better emer-gency communications. But the majority of the new expenditures have only one use rather than two or three. These include border security and port security (these may reduce drug smuggling but that is not a disaster threat), surveillance of citizens and suspicious organizations, radiological detection devices and programs, a vast expansion of highly secure bioterrorism laboratories, the biohaz-ard suits and equipment spread across our rural counties, attempts to coordinate intelligence agencies, "human assets" for penetrating terrorist cells, and so on. (As noted, these expenditures, specific to terrorism and limited to them, appear to have been at the expense of responding to natural disasters.) In contrast, reducing the size of targets will reduce the consequences of all three disaster sources. An all-targets approach is more efficient than the department's all-hazards approach.

Though it is only tangential to the thesis of this book, it is worth putting the threat into perspective and reflecting on the degree of

threat from radical Islamic terrorism. One hesitates to make a prediction in this area but I think the chances of an attack that could kill one thousand or more and/or significantly harm our economy (say, more than the 2003 blackout did) is extremely low, so low as not to warrant the expenditures made to prevent it. Somewhat higher, but still fairly low, is an attack of the scope of the two London subway bombings in 2005 or the Madrid train station bombing in 2004, the planned explosions of gas stations on the East Coast, and early planning of a coordinated attack on jetliners flying to the United States (As of this writing there was no indication that the latter involved terrorists who had tested or even fabricated the liquid bombs.)

I think we have the most to fear from natural disasters, next industrial disasters, and the least from terrorist disasters. We now have sufficient awareness and primitive prevention devices to forestall attacks similar to the 9/11 attack. A complicated, large-scale attack like that is very vulnerable to even a modest degree of preparation on our part. The attack almost failed at several points where we are now more secure. The airlines are on greater alert (though their cargo is still not inspected), and the police are more likely to investigate traffic violations by suspicious persons; the air traffic controllers are more likely to halt takeoffs at the first sign of a hijacking, which might have prevented at least two of the four hijacked planes from departing, and so on. The other suspected terrorist attacks on our territory since 9/11 have been far less sophisticated, and have been foiled. The reconnaissance of the Brooklyn Bridge to see if the cables could be cut with an acetylene torch was truly inept, for example, and it was ordered by a top Al Qaeda leader. (Shrader 2005) The surprising thing is that unsophisticated attacks on railroad tank cars, chemical storage tanks, spent-fuel pools at nuclear plants, and our milk supply have not occurred. The explanation may be that there are no or few active cells; the FBI recently declared that it could find no evidence of Al Qaeda cells in the United States, though other intelligence agencies appear to disagree. (Ross 2005)

We have not had a terror attack here in five years, despite the

ease with which one seemingly could occur. If we do have one, it is likely to be less than catastrophic, such as a subway bombing as in London or a truck bomb at a target with more symbolic than lethal consequences. Margaret Kosal, a chemist, however points out that small-scale chemical weapons (improvised chemical devices, or ICDs) are easy to manufacture, and terrorists' skills at improving their improvised explosive devices (IEDs) in Iraq are consistent with the possibility of making deadly ICDs. Chemical threats are underestimated, she argues. (Kosal 2005) I agree, but without a great deal of sophistication on the part of their makers, evidence of which is lacking, an improvised chemical device is not going to do *massive* damage.

However, a biological attack would be massive in its devastation. Scientific advances in just the last five years have made it possible for a small laboratory to download recipes (the genetic code) from the Internet for lethal synthetic viruses, order online from hundreds of eager suppliers the tiny bits of viral DNA necessary, and assemble them in a lab. A biochemist at Rutgers University is quoted as saying "It would be possible—fully legal—for a person to produce full-length 1918 influenza virus or Ebola virus genomes, along with kits containing detailed procedures and all other materials for reconstitution. It is also possible to advertise and to sell the product, in the United States or overseas." (Warrick 2006)

Experts say there are no adequate countermeasures, including stockpiling of vaccines for a multitude of unknown pathogens that can be set loose. According to CIA study in 2003, "the effects of some of these engineered biological agents could be worse than any disease known to man." Still, the capability of terrorists to launch such an attack is "judged to be low," according to a DHS intelligence officer. A more likely source is a lone scientist or biological hacker, such as the Unabomber or the person(s) who sent anthrax through the mail. (Warrick 2006)

Regardless of ICDs, truck bombs, biological agents that could destroy Islam itself, and so on, Islamic terrorists would do better to increase their attacks on authoritarian and failed Muslim states, such as Egypt, Kuwait, and especially Saudi Arabia, in the hope of

destabilizing their regimes. Further attacks on the United States would mainly be morale boosters, and given the success of both local and international terrorists in Iraq, and the mounting hatred of the United States in that part of the world, a morale boost is hardly needed. A truly horrendous attack on the U.S. soil might be counterproductive if it brought redoubled efforts from the West to attack terrorists. It is even reported that one Al Qaeda leader reprimanded bin Laden for the 9/11 attack since it led to the destruction of their training grounds. (Diamond 2006) (But another terrorist thought they should have shipped radioactive materials in the cargo of the hijacked planes!) (Whitlock 2006) Bin Laden was reported to have been quite surprised by the amount of damage done in that attack. Another massive attack on the scale of 9/11 may simply not be in their interests, aside from the difficulty of carrying it out, given our much heightened awareness and somewhat heightened defenses.

Both chemical and biological attacks would require more sophistication than any of the attacks that have occurred in recent years, and there is no evidence Al Qaeda has that capability. Milton Leitenberg argues that U.S. government officials and the media have exaggerated the threat of bioweapons, especially by nonstate actors. He criticizes the large expansion of biodefense research and development for two reasons. It is disproportionately large compared with that being spent in other public health areas. Second, the expanded research programs are likely to increase the wrong kind of interest in biological weapons and, in consequence, promote their proliferation. (Leitenberg 2004) There is no evidence of capability on the part of nonstate terrorist groups. The title of one Associated Press review is "Terrorists inept at waging war with chemicals." (Hanley 2005)

The Madrid train attack required sophisticated coordination and fairly sophisticated timing devices, but the explosive power cannot be described as massive. The London subway attacks were truly primitive and did not do the damage a more sophisticated attack could have. (This surprises me; there are much more devastating targets in these cities that would require only easily available

explosives and very little planning and, instead of killing one hundred or so in a subway, could kill thousands.) Though captured documents and interrogations have demonstrated Al Qaeda's interest in attacking nuclear power plants, chemical plants, and the electric power grid, none of these attacks have occurred. (A domestic white supremacist is serving eleven years in prison for possession of a half a million rounds of ammunition, sixty pipe bombs, remote-controlled bombs disguised as briefcases, and improvised hydrogen cyanide dispersal devices that could kill thousands in a minute. [Kosal 2005] It is possible that we have more to fear from even inept domestic terrorists than from international ones. We know they are here, and they have tried.) The all-targets approach I am recommending—reduce the size of targets—rather than the all-hazards approach of the DHS, which amounts to a disproportionate emphasis on terrorist threats, would reduce our vulnerability to all three of the disaster sources that we have been considering. Even if international (and domestic) terrorism were no longer a threat, we should reduce our vulnerability to natural and industrial disasters. If terrorism remains a significant threat, the consequence of an attack would be lessened by this reduction, and there would be fewer attractive targets as well as less-consequential ones.

## WHAT IS TO BE DONE?

Reducing the size of targets involves two areas of concern: population concentrations in risky areas, and parts of our critical infrastructure such as power, telecommunications, chemicals, and transportation. The federal government plays a role in the first through its programs of insurance and disaster relief and some federal standards, but much of this risky-areas problem is a state and local problem. However, the parts of the critical infrastructure that is our second area of concern are national in scope and thus are primarily a federal problem. Unfortunately, however, almost all of these infrastructures are in the hands of private corporations. While citizens have a say in electing federal, state, and local repre-

sentatives they have none in the case of organizations that control our infrastructure. We have to depend on regulation and market structure in this area. This in turn depends on our three branches of government: executive, legislative, and judicial. Our government, in turn, depends on our electoral system, which is seriously flawed. Every attempt to reduce our vulnerabilities will be compromised by our flawed electoral system.

## The Electoral System

By design, our forefathers ensured that the Senate would be unrepresentative, with the small states having the same power as the large ones. (It is a form of gerrymandering writ large, but at least senators run on a statewide basis, unlike the gerrymandered House, which enables its members to have safe seats, which in turn makes national issues less relevant.) We saw the effects of our unrepresentative Senate when small states received homeland-security funding grossly disproportionate to both their population size and their vulnerability to terrorist attacks. (Our electoral college system is a comparatively minor defect, although it can be a major defect if presidential elections continue to be close.) Equally serious is the transformation of our electoral system occasioned by the powerful impact of television campaign ads on viewers. Before the advent of television, candidates campaigned primarily through personal appearances and, to a limited extent, on the radio. Campaign financing did not require the enormous budget that TV does. Business and industry contributed to the campaigns, but so did individuals and groups not associated with business. Corruption in the form of buying privileged access to politicians and buying their votes existed, of course; it has always been a problem, peaking in the period from about 1870 to 1929. But the blatant cases were few and could be brought to light, and the lack of a huge war chest did not prevent politicians who were free of business blandishments from getting elected.

Now politicians spend most of their time raising money for the

all-important television ads that make or break most elections.
Even where there are safe seats for a party because of gerrymandering, a Democratic challenger in a Democratic district will need
campaign financing from business for his or her challenge. This has
given business and industry an enormous advantage over other interested parties because of their deep pockets. They are able to ensure that politicians amenable to their interests get nominated and
elected; as a consequence, the politicians have been able, as we
have seen, to weaken environmental rules, avoid building safer levees, allow settlements in risky areas, reduce safety inspections by
the Nuclear Regulatory Commission, allow Senator Gramm to institute the deregulation that led to the Enron scandal and concentration in the power industry, and direct homeland-security funds
to very questionable projects that benefit business. The larger the
corporation, the greater the political and economic power it is
likely to wield. Of course, this influence is even more striking in
campaigns for the presidency.

Campaign finance reform has been on the agenda for two decades, but very little progress has been made. Even the scandals regarding lobbying in 2005 and 2006 have not produced significant
changes in lobbying restrictions—another loophole in our democratic system. Full public financing of campaigns, or at least serious spending limits, is probably the most important reform that
could lead to the changes needed to reduce our vulnerability that I
have been advocating. (There is a possibility, however, that the Internet may reduce the power of corporate financing of political
campaigns. The political parties are finding that advertisements
on the Web and political bloggers are reaching the approximately
fifty million people who use the Web to get their political news.
These methods are much cheaper than TV ads and so might reduce
politicians' dependency on business financing, but only somewhat.
ISP concentration in the Internet might reduce this grass-roots
effort. Furthermore, it is possible that the critical swing voters
are unlikely to get their political news from the Internet, but from
Fox TV.)

## Trends In The Executive, Legislative, and Judicial Systems

The White House and Congress are responsible for our judicial system, and the role of Supreme Court justices and justices of the circuit courts is vital to any reforms I have been advocating. The trend in the courts has been to favor private property rights and rights of eminent domain, which favor business interests over community rights. This has hampered the states in their attempts to preserve wetlands, restrict settlements in risky areas, and control standards of construction. Most alarmingly, the Supreme Court has also asserted the right of federal regulations to preempt state regulations even when the federal standards are lower than the state ones. For example, preemptive federal regulations regarding industrial hazards are weaker than those of some states, so the lower federal standards hold.

Unfortunately, the courts are just following Congress here. Since the 2001 election of President Bush and the dominance of Republicans first in one, then in both Houses of Congress, there has been a systematic and marked preemption of states' laws and regulation in social policies; health, safety, and environment protection; and consumer protection laws. A study ordered by Representative Henry Waxman (D-CA) indicated that in the past five years, Congress has voted fifty-seven times to preempt state laws and regulations, passing twenty-seven preemption laws. The federal floor, below which states cannot fall, is replaced by a federal ceiling, below which they must operate. This prevented states from setting higher standards than the legislated federal ones on air pollution, health insurance and health safety, contaminated food, spam prevention, and gun control and gave immunity from liability to vaccine manufacturers, gun manufacturers and dealers, rental car agencies, manufacturers of homeland-security products, fast-food restaurants, and manufactures of dietary supplements. The report quotes a *Los Angeles Times* story: "the Bush administration is providing industries with an unprecedented degree of protection at the

expense of an individual's right to sue and a state's right to regulate." (U.S. Congress 2006) Of particular concern to us is the Energy Policy Act of 2005, which "strips states of their authority over siting of liquefied natural gas terminals, notwithstanding the significant public safety and environmental concerns associated with the construction and operation of these facilities." Pending is a bill to limit states' authority to regulate toxic substances. (U.S. Congress 2006)

The country has taken small steps to reduce population concentrations in risky areas. FEMA relocated some communities following disastrous floods, turning the land into water-absorbing wetlands, but the program was discontinued under President George W. Bush (ironically on the same day a destructive earthquake demonstrated the program's value). The program could be restored and expanded. We could make a more vigorous attempt under a reorganized FEMA that began to recognize basic vulnerabilities. Unfortunately, as we saw in Missouri, local interests have succeeded in rebuilding on floodplains following disastrous floods, still with inadequate levy protection. After the 1993 Mississippi River flood, Missouri passed legislation that prohibited any county in the state from setting higher standards than the state, whose standards were already low. This is an instance where federal regulation should supersede state and local laws, unless local ones were more stringent. With more recognition of how the rest of the nation eventually helps pay for this shortsightedness, national legislation might have a chance.

Unfortunately our Congress, the Bush administration, and our Supreme Court have all joined together to weaken wetlands preservation since 2001. In Holland, the shock of a 1954 storm as large as Katrina led its government to take strong measures to protect its citizens. These eventually included not just the dikes and other forms of protection but deconcentration of especially risky areas and the restoration of wetlands—entire communities were moved. Neither the U.S. Congress nor the White House has been willing to fund the restoration of the wetlands on the Gulf Coast, which would go a bit of the way to making New Orleans a viable city

(downsizing is needed in addition) and protect many others along the Gulf Coast. All the necessary measures are far from impossible: relocation, enforcement of standards, reduction of federal flood assistance where business and citizens do not take out insurance or violate standards, repair of levees, and restoration of wetlands. Some of these address prevention and remediation rather than the basic vulnerabilities that are the major concern of this book, but increased awareness of the need for prevention and remediation could lead to more awareness of the need to reduce basic vulnerabilities. A credible announcement that the federal government would refuse subsidized flood insurance for vulnerable areas and would provide no relief in the event of a disaster that could be averted by population deconcentration would do wonders for long-term planning, building standards, evacuation routes, and population densities in risky areas such as the Gulf Coast and the southeastern seaboard.

Extensive federal regulation governs the use and handling of highly hazardous materials, particularly chemicals. But there appears to be no significant legislation restricting the volume of storage of such materials or their location. The most promising legislation regarding the protection of chemical plants from terrorists in recent years focused on physical protection of chemical plants and other users of toxic substances, and requires only consideration of safer processes, but nothing on storage volumes or processing volumes. The law, sponsored by then senator Corzine, did not make it out of committee. A weaker one was still under consideration in March 2006, but it contains the pernicious clause that would require states such as New Jersey, with higher safety standards than the federal ones, to lower them. Given the failure of the earlier legislation and limitations of the current proposal, I am not optimistic about the more basic approach of reducing large concentrations of hazardous materials in storage and in production processes, but it is not inconceivable. What probably is inconceivable, unfortunately, is a breakup of those corporations vital to our critical infrastructure into networks of smaller firms where, I believe, we are much more likely to find reliability and security.

We have successful regulations and incentives in many other areas of our lives. Federal government standards (even though not well enforced) govern transportation, civil rights, education, worker protection, health care, and so on. The reduction in toxic emissions since the establishment of the Environmental Protection Agency is certainly encouraging. Unfortunately, as we have seen, the regulations have emphasized voluntary adoption of standards and un-certified inspections. The result has been that the firms that signed on to the standards reduce their emissions less than those that do not, presumably because they expect to escape scrutiny. But without regulations, especially those from the federal government, our society would obviously be much more vulnerable to our three sources of disaster. Many of these regulations were once stronger than they are now, especially in the antitrust area and pollution abatement. The ideology prevailing in our federal court system does not favor tougher regulations.

The Supreme Court and the Circuit Courts are not immune to public sentiment. But those sentiments must be very, very strong. At best, strong public sentiment affects specific issues rather than the prevailing ideology. The Supreme Court and the circuit courts are much more responsive to the politics of the political party that put them in power, by virtue of presidential appointments of judges to these positions and consent to these nominations by the party controlling Congress. Better regulations will have to stem from the electoral system, and even then, the process will take time because of the lifetime appointments of circuit and Supreme Court justices.

One alternative to government regulation would be a change in court rulings that would make it easier for citizens to sue corporations. Citizens have had a great deal of difficulty in suing corporations for damage done to themselves or to their environment, while recent rulings have made it easier for citizens and special interest groups to sue the government in cases of "private property rights." But we should not forget that it is only in the past seventy-five years or so that citizens have had any significant chance to sue powerful organizations. Citizens' rights to sue corporations expanded in the 1960s through the 1980s, partly in association with the environ-

mental movement that was discussed in the chapter on the chemical industry. For a time, the courts went along with this. But since then, there seems to have been a contraction. Were a reinvigorated environmental movement to embrace the analysis presented in this book, there might be some success in this direction.

What is required is the passage of specific laws that would establish violations that then could form the basis for citizen's legal action. If the laws were upheld by the courts, and this is problematic, organizations would seek insurance policies that would protect them from lawsuits. Even without specific laws, we have seen the important role that liability suits might play in promoting a more reliable and more secure Internet. In addition, as the principal customer of operating systems and software, the federal government could require more reliable and secure performance than we are getting, that is, changing priorities of the suppliers. Again, it is a state—Massachusetts—that is leading the way. It is even conceivable that in many parts of our critical infrastructure, an insurance company would reduce a firm's premiums if it made the necessary expenditures to ensure access to multiple suppliers and multiple customers, so as to prevent the kind of breakdowns that excessive dependencies and critical nodes can cause.

(Insurance companies obviously play an important role in assessing and reducing risks. But they have no incentive to do so if the costs of disasters are fully socialized, that is, borne by the public at large, as is the case with most natural disasters. They are unlikely to bring to our attention the widespread negative incentives of making the public pay, since then they might have to pay. Governmental regulation could require it, but as we have seen in the examples of rebuilding on floodplains after the 1993 Mississippi River flood, hurricanes repeatedly destroying homes on a barrier island and the government paying handsomely to rebuild them, and the protection afforded nuclear power plants by the Price-Anderson Act, it is unlikely to.)

Nevertheless, despite all these obstacles, corporations are far more vulnerable to lawsuits by citizens and voluntary associations than they were one hundred or even fifty years ago. There is, then,

some reason to hope that the upward trend of the past could be restored. One is entitled to hope that the present state of affairs is only a pause or a small setback in a historical trend that clearly favors federal involvement in disaster relief, safety standards and safety regulations, antitrust legislation, and the closer regulation of toxic and explosive substances. Laws that could protect our critical infrastructure have been weakened, or not enforced, for approximately two decades. This process was accelerated under the present administration, but it did not begin there. Nevertheless it could be reversed.

### Could Further Calamities Help?

Will further calamity capture our attention sufficiently to change our direction? The experience of Hurricane Katrina is not encouraging. We may stick some fingers in the dikes and shout warnings, but like the bungled recovery effort still going on as of summer 2006, further engineering of the levees will probably also be bungled (even private contractors doing the work warned the Army Corps of Engineers that what they were doing was faulty), the Gulf states will not tax themselves to raise money for restoration of the natural barriers (the federal government has offered only a pittance), the mayor of New Orleans has declared that every one should come back, which would produce a population concentration that cannot be defended. Our response to 9/11, establishing the Department of Homeland Security, was a disaster in its own right because despite our huge expenditures on safety, it has done very little to thwart terrorism but a great deal to further pork-barreling in government, further the privatization of the federal government, de-unionize a significant portion of its employees, limit civil liberties, and help justify a disastrous war in Iraq. We may hope that another large terrorist attack does not occur, not just because of the immediate damage to lives and property, but because it may further reduce our ability to handle the more frequent and predictable natural and industrial disasters, provide more op-

portunities for the looting of the public treasury by private firms, further limit civil liberties, and even launch a new war. We should not expect further "wake up calls" from terrorists to make us safer.

Unfortunately it may take a disastrous nuclear power plant accident with extensive radiation over a few hundred square miles to weaken control of poorly regulated private utilities over the dangerous nuclear beast. A revitalized Nuclear Regulatory Commission could do the job if senators heavily financed by public utility firms were unable to threaten the agency with budget cuts unless it cut back on inspections, as noted in chapter 5. It is encouraging that a Connecticut public utility agency persisted in exposing the malfeasance of the executives that brought the Millstone nuclear power plant so close to a serious accident and gave us our first clear documentation of executive failure in one of our large utilities. Another grim wake-up call that might produce legislation would be a massive contamination of a 400,000-gallon milk storage silo that would tragically endanger thousands of people, especially children. This just might prompt federal legislation not only to make them safer, the inevitable first response, but to deconcentrate this industry, the proper, long-term second response.

The spread of mad cow disease through our giant feedlots might do the same for the cattle industry, which is highly centralized. (But even the voluntary inspection plan was recently severely cut back. It was already flawed because the industry would only inspect those dead animals that would not be likely to have the disease.) (McNeil 2006) A few chemical plant explosions might wake up Congress, but the run of mine disasters in 2006 has prompted very little reform. We can expect more massive blackouts, but it is unlikely that their connection to deregulation will be seriously investigated. As we have noted, the 2003 blackout has failed to bring about such an investigation. It is possible that hackers (or even terrorists, though that is unlikely) could cause such a wrenching disablement of the Internet through malware planted in Microsoft's software that we might see the appearance of liability lawsuits. (It would take a change in attitude on the part of the Supreme Court, but twenty years ago the Court was more willing to entertain the

notion of corporate liability for harms done to citizens; it could go back to that position if there were huge public pressure.) The specter of very expensive liability insurance for firms that pay insufficient attention to reliability and security would catch everyone's attention. This might delay the next version of an operating system such as Microsoft's Vista by two or three years, but it would produce a more reliable and a more secure one. I think most of us could wait that long.

The only hopeful signs are, unfortunately, tiny. But there are beginnings. We might get a marginally reasonable chemical safety bill out of the Republican-controlled subcommittee, and if it did come out it is quite likely that it could pass even in the Congress of 2006. (The choke points for preventing enlightened legislation are often as small as the composition of a subcommittee.) A variety of national engineering and environmental groups have put forth realistic plans for restoring the wetlands in the Gulf Coast. Most people were not aware of the basic cause of New Orleans' drowning, but they are now, and support for these plans could mount. Though protecting the Gulf is a federal problem, local education, and even local education of those not living in the Gulf states, can create an issue that congressional representatives might have to attend to.

One important thing we need to do is assess our vulnerabilities. The comical effort of the DHS to do this is scandalous—allowing states to declare petting zoos and flea markets as terrorist targets. (Lipton 2006b) More important, the assessment ignores the targets of natural disasters and industrial disasters. Half in jest, I would recommend the disassembly of the twenty-two agencies thrown together in the Department of Homeland Security and the formation of a Department of Homeland Vulnerability instead. The GAO could tell us what the most dysfunctional relocations of the twenty-two agencies were, so we could return them to their former status as independent agencies, and which agency relocations were useful and should be preserved. An independent commission, with engineering experts such as Ted Lewis and wise men such as Circuit Court Justice Richard Posner could survey our basic vulnera-

bilities and make highly publicized recommendations to Congress to reduce them. (Lewis 2006; Posner 2004)

Even without cataclysmic prods to our federal government, including its courts, it is still possible that incremental changes could be sufficient to protect us. It took decades of slowly evolving court decisions in the nineteenth century to finally stop blaming workers who were killed on the job for their deaths. Our federal Food and Drug Administration is the result of an evolutionary process rather than of a single catastrophe. Almost no one had health insurance in my childhood and we came very close to having national health insurance, such as all other industrialized nations have, during the Truman administration. Business interests, which helped defeat that attempt, now have reason to regret their stand. Steady pressure from unions and public interest groups produced the Occupational Safety and Health Administration. There are some federal building standards; there could be more. We are better prepared today to handle the disasters experienced in the past. However, there are more disasters today than in the past, and they are more serious. We are more prepared, and yet we can improve. But we have hardly began to do the most effective thing: reducing the size of the targets that inevitably will be attacked. We are not safe. Nor can we ever be fully safe, for nature, organizations, and terrorists promise that we will have disasters evermore. Let us minimize their consequences by minimizing the size of our vulnerable targets.

# Appendix A

## Three Types of Redundancy

----------------------------

ONE THING THAT CHARACTERIZES the Internet, and to some extent perhaps the power grid, are the levels of redundancy. This is a tricky concept. I have held that redundancy that is designed in, and thus designers are aware of it, is essential and safe, but that redundancies that are added later, as safety devices, are often the source of system failures. (Perrow 1999) Their potential interaction with other failures is not foreseen by designers and not comprehensible to operators. Scott Sagan, however, goes further and persuasively argues that even designed in redundancies are a major source of accidents. He applies this to the nuclear power plant security issue. (Sagan 1996; Sagan 2003) Aside from increasing the interactive complexity of the system—always a risk—redundancies produce a false sense of security. Tentatively, I would propose that this is true for most systems, but not for heavily networked systems. In these, nodes can be connected through multiple, and thus redundant, paths; new paths can grow quickly; and malfunctioning units that could threaten the whole can be quickly, even instantaneously, isolated, severed from the Web, and, if desired, only temporarily. This is true of both the power grids and the Internet.

Furthermore, there is more than one type of redundancy. (I am indebted to Allesandro Narduzzo of the University of Bologna for suggesting the two additional types of redundancy and providing

me with the following map of the Internet. The responsibility for errors, however, is solely mine) We normally think of two identical components, one being redundant and only used if the other fails, or two different power sources, the usual connection to the power grid and a standby diesel generator. This could be called *replacement redundancy*. But the Internet appears to have redundancy in two additional forms, involving the replication of two parts of the system's structure.

A computer sends a message that includes the address (tier four, the lowest). It goes to an Internet router, tier three, which notes the address, and sends it to a local service provider, such as a university, corporation, or commercial service provider (tier two). At tier one, the top level, called the Internet "backbone," the tier two service providers are linked. This is a hierarchical structure, since there are four levels, but given its enormous size no other system could get by with only four levels. Furthermore, growth of the system is not produced by adding levels of hierarchy, as happens in most organizations, in order to direct and coordinate and service the additional users. Instead, using the advantages of distributed systems, growth is achieved by replicating the three tiers above the final user and introducing links between the nodes within each level, called *peering links*. Servers at tier two and routers at tier three can be added, but no new levels.

Duplicating the connections between peer routers at the lowest coordinating and routing level does two things: it provides greater point-to-point available bandwidth—multiplying the paths—but it also introduces greater reliability, though only as a secondary consequence of adding new potential paths. This reliability is not achieved through replacement redundancy, since all the paths are always active and available to all the users, but through redundancy by replicating the contents of servers and routers. Let us call this form *replication/link redundancy,* to capture both the replication aspect and to capture the multiple links available between "peers." The link redundancy feature is still present in the RAID feature, the redundant arrays of inexpensive disks (sometimes called redundant array of independent drives), in case the main system

breaks down. But the reliability of link redundancy emerges as a property of a *distributed system* (semi-independent self-activating units or functions that are distributed throughout the system, capable of performing tasks that otherwise would be restricted to a central unit). The architecture is replicated in units that are always active. But even more important than the replication of the architecture, allowing multiple pathways for packets to travel, is the replication of the packets and their addresses.

This constitutes an additional form of redundancy beyond replacement redundancy. But there is a third form. "Link redundancy," Narduzzo writes, "is the most evident effect (especially because it is analogous to other systems, such as power grids, which are based upon connections), but it is not the most critical one. By replicating its architecture, the Internet also duplicates some information, in particular, the memory of its own structure." (Personal communication) This is called mirroring, so we will label it *mirroring redundancy*. It goes beyond duplicating connections—duplicating the architecture—because the *content* is mirrored, rather than just the connection. For example, messages are broken up into many packets which are sent by different routes by routers. If one packet is lost along the way, the servers that reconstitute the message will know a packet is missing because the "header" of any packet that arrives has the address of all the other packets—the mirroring aspect—and this will show which packets are missing, and where they came from. The server will tell the computer that sent the packets to send this one again.

Domain name system (DNS) servers are essentially databases that store the updated lists of Internet addresses ("IP numbers") that the routers need to have. This information is replicated at each node of the network. Two DNS servers may be serving two quite separate multiple streams of data that are physically independent. But they can mirror each other such that their internal states are identical; they both have all the addresses. This mirror redundancy means that if one server is disrupted, or a packet it is serving is faulty, the data streams to it can be processed by the other server without error, and virtually instantaneously.

The mirrored servers and the DNS servers are always active and are physically dispersed in the network and can be accessed from the closest site and thus be more quickly available. When a part of the network fails, as it often momentarily does, the mirrored information is still available, even if it takes a bit more time to find it. Reliability is a secondary effect, the primary purpose of this aspect of the architecture is speed—distributed information is more quickly accessed than centralized information.

These two forms of redundancy, replication/link, and mirroring, do not appear to introduce the hazards that I and especially Sagan have detected in the normal form of replacement redundancy. They are always online, and thus do not decay or need testing; they do not give the operator a false sense of security or induce inattentiveness; and it may be that the additional complexity they contribute to the system is not a source of unexpected interaction of failures. (One troubling aspect of this system, however, has already been noted: There is a subtle additional hierarchy in the Internet— the thirteen root servers and the ten top-level domain servers. A hacker attack, perhaps by a terrorist, on these that was timed to coincide with an attack on another infrastructure, such as telephone service, a banking center, or explosive or toxic concentrations, could magnify the damage and prevent recovery. Network organizations are not immune to attack of this sort, despite the high reliability and flexibility they achieve while still being enormous in size and highly efficient.)

# Appendix B

## Networks of Small Firms

------------------------

THE QUESTION ARISES AS TO whether more reciprocity will be obtained in a system that has a network of firms versus a system that consists of just one firm, hierarchically organized. One could argue that the successful single firm that has brought within it most of the components needed will have a high degree of reciprocity among the divisions, departments, and work groups. But such a setup is unlikely to have multiple suppliers and each component has only one customer, so there is a loss in resiliency. On the other hand, one might say that the competition among a variety of networked firms might inhibit reciprocity. However, the literature stresses the sense of common fate and sense of community in these networks, especially if the units are in the same geographical region. In contrast to the literature on networks, the literature on the problems of organizing in a single firm stresses the lack of reciprocity. In the principal-agent literature that is so dominant today, the emphasis is on the agent "shirking" responsibilities and the necessity for authoritative leadership and surveillance to counteract this shirking. The classic statement is from Oliver Williamson (1975) and challenged by myself. (Perrow 1981; Perrow 1986) The literature on networks of small firms, on the other hand, stresses the degree of reciprocity.

The reliability of networks of small firms is more difficult to

asses, since there is no convenient metric. But students of small-firm networks attest to their robustness, even in the face of attempted consolidations by large organizations. Saxenian effectively contrasts the decline of the nonnetworked group of high-technology firms around Boston's Route 128, when federal funding declined and Japanese mass-production techniques matured, with the networks of small firms in California's Silicon Valley, who charged forward with new innovations for new markets. (Saxenian 1996) Despite predictions of their imminent demise (Harrison 1994; Harrison 1999), dating back to the 1980s when they were first discovered and theorized, the small-firm networks of Northern Italy have survived and dominate a variety of low- and high-tech industries. In the United States, the highly networked biotech firms are prospering, for the time, despite their linkages with the huge pharmaceutical firms. (Powell, Koput, and Smith-Doerr 1996) Particular firms in small-firm networks come and go, but the employees and their skills stay, moving from one firm to another as technologies, products, and markets change.

Some economists argue that networks of small firms are inefficient because of increased transaction costs—everyone is buying or selling with everyone else—but elsewhere I have argued that transactions are not necessarily increased in small-firm networks, they are just external to each firm rather than occurring within it. Problems of self-interest with guile are reduced since such networks generate trust, and they also have lower surveillance and monitoring costs. (Perrow 1981; Perrow 1986; Perrow 1992) The favored design of large, multidivisional firms with forward and backward integration need not reduce dependencies. The dependency is merely transferred from the environment to the system, and alternative customers or suppliers are less available, reducing safety. Adaptive multiuse systems and decentralized networked systems are probably the most resilient to natural, industrial, and deliberate disasters. But our government, our economic ideology, our economists, and the economic interests of the large players do not favor such systems. Consequently, vulnerabilities are increased.

Increasing the size of an organization increases hierarchy and

centralized control. In nineteenth-century America, for example, economic output expanded greatly without mass production, and without vertical integration (incorporating supply and distribution services, legal services, construction, and so on into the organization rather than purchasing it from others). Instead, a firm added more shops, or work groups, which replicated those already there. According to the standard functionalist argument, the innovation that mass production produced was to separate out the distinctive tasks in each shop (e.g., welding), merge them into specialized functions, and link them together in a production sequence, ideally into an assembly line. This also required that the goods and services heretofore purchased in the market be produced within the firm in order to ensure complete control and complete reliability of these goods and services. Vertical integration was driven by efficiency, in this literature.

I have argued elsewhere that the source of vertical integration that occurred suddenly at the end of the nineteenth century was only in small part efficiency; much of it, and the most fateful aspects of vertical integration, was a search for market control and the concentration of economic and political power. The bulk of economic enterprises could have remained decentralized and small or medium-sized without significant economic inefficiencies, and have produced substantial social efficiencies. (Perrow 2002) There were undoubted increases in profits with mass production, and these were socially acceptable because the social externalities that came with market control, while appearing quickly and alarmingly in some industries such as the railroads and petroleum and steel, were slow to appear in the rest of the economy. There was also less interdependency in the economy and society, making mass production in a few industries less consequential. Now that has changed, and scale appears to have many undesirable features, some of them unanticipated until the very end of the twentieth century.

The multidivisional form was in part an attempt at decentralization. But multidivisional firms often fail at buffering their divisions from each other because they seek coupling that is too tight—for example, imposing the same organizational structure or accounting

system or personnel policies on units with diverse technologies that need diverse structures. The successful firms are loosely linked stand-alone systems that allow different technologies to have different organizational structures. There may be five or ten divisions, some centralized, some decentralized, having a variety of "cultures," but with a common headquarters unit. Better yet, if the technologies and environments are truly diverse, they will be linked only by equity ownership. (For the classic statement on forms of linkages see Powell 1990.)

Not only do we have large vertically integrated corporations and multidivisional firms that still remains centralized, but the global economy has produced another form of concentration. It threatens the networks of small firms that were doing so well in the late twentieth century in northern Italy, northern Europe, Japan, and even parts of the United States. First, many of our consumer goods came to be produced in offshore factories, but initially these were relatively small. However, the last twenty years has seen the emergence of giant retailers such as Wal-Mart and, even above them in the commodity chain, the rise of giant transnational contractors based primarily in Hong Kong, Taiwan, South Korea, and China. They operate large factories in the less-developed world and provide a potential counterweight to the growing power of retailers. (Applebaum 2005) The disaster threat, other than to our economy, is not apparent, but the economic concentration in consumer goods could spread to products more critical to our critical infrastructure. For example, economic concentration in the management of our ports has led to the ability of a foreign nation that was once quite hospitable to terrorists, and still the nexus of the arms trade, to almost take over the principal port management system in the United States. The dangers of this tight integration and dependency on single suppliers is amply demonstrated by Barry Lynn. (Lynn 2005)

# Bibliography

----------------------------

Administration, Energy Information. 2000a. "The changing structure of the electric power industry 2000: An update," October. Washington, DC: U.S. Department of Energy. http://www.eia.doe.gov/cneaf/electricity/chg_stru_update/update2000.pdf.

———. 2000b. "Upgrading transmission capacity for wholesale electric power trade." Washington, DC: U.S. Department of Energy. http://www.eia.doe.gov/cneaf/pubs_html/feat_trans_capacity/w_sale.html.

———. 2003. "Electric power annual." Washington, DC: U.S. Department of Energy. http://www.eia.doe.gov/cneaf/electricity/epa/epa_sum.html.

———. 2005. "Average retail price of electricity to ultimate customers by end-use sector," November. Washington, DC: U.S. Department of Energy. http://www.eia.doe.gov/cneaf/electricity/epa/epat7p4.html.

AFL-CIO. 2006. "Unchecked: How Wal-Mart uses its might to block port security." *AFL-CIO*, April (accessed April 12, 2006). http://www.aflcio.org/corporatewatch/walmart/upload/walmart_unchecked_0406.pdf.

Albert, Reka, Hawoong Jeong, and Albert-Laszo Barabasi. 2000. "Error and attack tolerance of complex networks." *Nature* 406: 378–382.

Allen, Barbara L. 2003. *Uneasy alchemy: Citizens and experts in Louisiana's chemical corridor disputes.* Cambridge: MIT Press.

Allenby, Brad, and Jonathan Fink. 2005. "Toward inherently secure and resilient societies." *Science* 309 (August 12):1034–1036.

Altman, Lawrence K., and Gina Kolata. 2002. "Anthrax missteps offer guide to fight next bioterror battle." *New York Times,* January 6.

Alverez, Robert. 2002. "What about the spent fuel?" *Bulletin of the Atomic Scientists* 58: 45 – 51.

American Red Cross. 2005. "Hurricane Katrina: Why is the Red Cross not in New Orleans?" http://www.redcross.org/faq/0,1096,0_682_4524,00.html.

Amin, Ash. 2000. "Industrial districts." In *A Companion to Economic Geography*, edited by E. Sheppard and T. J. Barnes. Oxford, UK: Blackwell.

Amin, Massoud. 2000. "National infrastructures as complex interactive networks." In *Automation, control, and complexity: An integrated approach*, edited by T. Samad and J. Weyrauch. New York: Wiley.

———. 2001. "Toward self-healing energy infrastructure systems." *IEEE Computer Applications in Power* January, 20–28.

Anonymous [Michael Scheuer]. 2004. Imperial hubris: Why the West is losing the war on terror. Washington, DC: Brassey's.

Appelbaum, Richard P., and Nelson Lichtenstein. 2006. "A new world of retail supremacy: Supply-chains and workers' chains in the age of Wal-Mart." *International Labor and Working Class History* 70 (forthcoming).

Appelbaum, Richard P., and William I. Robinson. 2005. "Toward a critical globalization studies: Continued debates, new directions, neglected topics." In *Critical Globalization Studies*, edited by Richard P. Appelbaum and William I. Robinson, ix–xxxiii New York: Routledge.

Ashford, Nicholas A., et al. 1993. "The encouragement of technological change for preventing chemical accidents: Moving from secondary prevention and mitigation to primary prevention," July. Center for technology, Policy and Industrial Development, MIT. https://dspace.mit.edu/handle/1721.1/1561.

Ashford, Nicholas A., and R. F. Stone. 1991. "Liability, innovation, and safety in the chemical industry." In *The liability maze: The impact of liability law on safety and innovation*, edited by P. W. Huber and R. E. Litan, 367–427. Washington, DC: Brookings Institution.

Baldauf, Scott. 2004. "Bhopal gas tragedy lives on, 20 years later." *Christian Science Monitor*, May 4.

Barr, Stephen. 2004. "Three unions come out against proposed Homeland Security personnel rules." *Washington Post*, March 23.

Bekker, Scott. 2005. "SQL 2005: The integrated stack is back." Redmondmag .com, December. http://redmondmag.com/features/article.asp?editorialsid=533.

Benjamin, Daniel, and Steven Simon. 2002. *The age of sacred terror*. New York: Random House.

Berenson, Alex. 2002. "Mystery of Enron and California's power crisis." *New York Times*, May 9.

Berger, Eric. 2001. "Keeping its head above water" *Houston Chronicle*, December 1.

Bier, Vicki. 2006. "Hurricane Katrina as a bureaucratic nightmare." In *On risk and disaster: Lessons from Hurricane Katrina*, edited by Ronald J. Daniels, Donald F. Kettl, and Howard Kunreuther, 243–254. Philadelpia: University of Pennsylvania Press.

Blair, Jayson. 2002. "Downtown phone network is vulnerable, report says." *New York Times*, August 6.

Block, Robert, and Russell Gold. 2005. "Managing a hurricane—differently." *Wall Street Journal*, September 23.

Blum, Justin. 2005. "Hackers target U.S. power grid." *Washington Post*, March 11.

Blumenfeld, Laura. 2003. "Former aide takes aim at war on terror." *Washington Post,* June 16.

Board, Chemical Safety. 2002. "DPC Enterprises chlorine release, Festus, Missouri, August 14, 2002." Washington, DC: Chemical Safety and Hazard Investigation Board. http://www.chemsafety.gov/.

———. 2004. "DPC Enterprises chlorine release," June 9. Washington, DC: Chemical Safety and Hazard Investigation Board. http://www.csb.gov/index .cfm?folder=current_investigations&page=info&INV_ID=45.

Borenstein, S. 2002. "Concern heightened over nuclear attacks." *San Jose Mercury News,* January 31.

Bourne, Joel K. Jr. 2004. "Gone with the water." *National Geographic,* October, 88–105. http://magma.nationalgeographic.com/ngm/0410/feature5/?fs=www7 .nationalgeographic.com.

Bradshaw, Larry, and Lorrie Beth Slonsky. 2005. "Hurricane Katrina—our experiences." EMS Network News. http://www.emsnetwork.org/artman/publish/ article_18427.shtml.

Bridges, Andrew. 2006. "Flood-zone development said to raise risk." *Boston Globe,* February 19. http://www.boston.com/news/nation/articles/2006/02/19/ flood_zone_development_said_to_raise_risk/.

Brzezinski, Matthew. 2004a. *Fortress America: An inside look at the coming surveillance state.* New York: Bantam Dell.

———. 2004b. "Red alert." *Mother Jones,* September/October, 38–43, 92, 94. www.motherjones.com/cgi-bin/print_article.pl?.

Bullard, R.D. 1990. *Dumping in Dixie: Race, class, and environmental quality.* Boulder: Westview.

Burdeau, Cain. 2005. "Katrina turns oil-contaminated town into epitome of despair." Associated Press, November 15. http://web.lexis-nexis.com/universe/ document?_m=6fccfeba2ca6176db71fc12eb398991c&_docnum=16&wchp =dGLbVlz-zSkVb&_md5=54cdc992202e1dc61d6854b23b3b7e3b.

Byrd, Sen. Robert C. 2004. "Senate Speech." Truthout, December 8. http://www .truthout.org/docs_04/121004X.shtml.

Cable, Josh. 2006. "FirstEnergy fined $28 million for hiding Davis-Besse damage from NRC." Occupational Hazards, January 20. http://www.occupational hazards.com/articles/14591.

Cappiello, Dina. 2005. "Hurricane aftermath; spills from storms staining the coast." *Houston Chronicle,* November 13.

Caro, Robert A. 2002. *Master of the Senate.* New York: Knopf.

Casazza, J. A. 1998. "Blackouts: Is the risk increasing?" *Electrical World* 212: 62–64.

Casazza, J. A., and F. Delea. 2003. *Understanding electric power systems: An overview of the technology and the marketplace.* New York: Wiley.

CBS News. 2004. "U.S. plants: Open to terrorists." CBS News, June 13.

Centre, Human Security. 2005. *The human security report 2005.* New York: Oxford.

Changnon, Stanley A. 1998. "The historical struggle with floods on the Mississippi River basin." *Water International* 23 (December) 263–271.

Changnon, Stanley A., and David Changnon. 1999. "Record-high losses for weather disasters in the United States during the 1990's: How excessive and why?" *Natural Hazards* 18:287–300.

Chemical Safety Board. *See* Board, Chemical Safety.

Churchill, Ward, and Jim Vander Wall. 2002. *The COINTELPRO papers: Documents from the FBI's secret war against dissent in the U.S.* Boston: South End.

Clarke, Lee. 1999. *Mission improbable.* Chicago: University of Chicago Press.

———. 2002. "Panic: Myth or reality?" *Contexts* 1 (Fall): 21–27.

———. 2005. *Worst cases: Terror and catastrophe in the popular imagination.* Chicago: University of Chicago Press.

Clarke, Lee, and Charles Perrow. 1996. "Prosaic organizational failure." *American Behavioral Scientist* 39 (August): 1040–1056.

Clarke, Richard A. 2004. *Against all enemies: Inside America's war on terror.* New York: Free Press.

Commission, Federal Energy Regulatory. 2005. "The commission's response to the California electricity crisis and timeline for distribution of refunds." December 27. Washington, DC: Federal Energy Regulatory Commission. http://www.ferc.gov/legal/staff-reports/comm-response.pdf#xml=http://search.atomz.com/search/pdfhelper.tk?sp-o=2,100000,0.

Copps, Michael J. 2003. "The beginning of the end of the Internet? Discrimination, closed networks, and the future of cyberspace." Lecture, New America Foundation, October 9. http://www.newamerica.net/Download_Docs/pdfs/Doc_File_194_1.pdf.

Council, National Research. 2001 *The Internet under crisis conditions: Learning from September 11.* Washington, DC: National Academies Press.

Dahl, Robert A. 2003. *How democratic is the American constitution?* New Haven, CT: Yale University Press.

Davis, Mike. 1990. *City of quartz: Excavating the future in Los Angeles.* London and New York: Verso.

Delmas, M. A. 2000. "Barriers and incentives to the adoption of ISO 14001 by firms in the United States." *Duke Environmental Law & Policy Forum.* 11: 1–38.

Delmas, M., and A. Keller. Forthcoming. "Strategic free riding in voluntary programs: The case of the US EPA Wastewise Program." *Policy Sciences.*

Diamond, John. 2006. "Captured Al-Qaeda documents show inner struggles, strategies." *USA Today,* February 16.

Donohue, Laura K. 2006. "You're being watched. . . ." *Los Angeles Times,* January 12. http://www.latimes.com/news/printedition/opinion/la-oe-donohue12jan12,0,1473176.story.

Drees, Caroline. 2004. "U.S. terrorism policy spawns steady staff exodus" Reuters, April 7.

Eagle, Steven J. Forthcoming. "Securing a reliable electricity grid." Working Paper in Law and Economics no. 05-31, George Mason University School of Law.

Egan, Timothy. 2005. "Tapes show Enron arranged plant shutdown." *New York Times,* February 4.

Eggan, Dan. 2004. "Measure expands police powers." *Washington Post*, December 10.

Elliston, Jon. 2004. "Disaster in the making." *Independent Weekly*, September 22. http://www.indyweek.com/durham/2004–09–22/cover.html.

Energy, U.S. Department of. 2002. "National transmission grid study." Washington, DC: U.S. Department of Energy.

Energy Information Administration. *See* Administration, Energy Information.

Erikson, Kai. 1994. *A new species of trouble: Explorations in disaster, trauma, and community*. New York: Norton.

Ervin, Clark Kent. 2004. "Mission: Difficult, but not impossible." *New York Times*, December 27.

———. 2006. *Open target: Where America is vulnerable to attack*. New York: Palgrave Macmillan.

Everest, Larry. 1985. *Behind the prison cloud: Union Carbide's Bhopal massacre*. Chicago: Banner.

Federal Energy Regulatory Commission. *See* Comission, Federal Energy Regulatory.

Flynn, Stephen. 2004. *America the vulnerable: How our government is failing to protect us from terrorism*. New York: HarperCollins.

Funk, John, and John Mangels. 2003. "Besse engineer says warnings ignored." *Cleveland Plain Dealer*, February 19.

Funk, John, Teresa Dixon Murray, and Tom Diemer. 2004. "Blackout report rips FirstEnergy." *Cleveland Plain Dealer*, April 6.

Funk, John, and Mike Tobin. 2006. "Three accused of lying about reactor." *Cleveland Plain Dealer*, January 20.

GAO. 2003 "Oversight of security at commercial nuclear power plants needs to be strengthened," August 24. Washington, DC: General Accounting Office. http://www.gao.gov/cgi-bin/getrpt?GAO-03–752.

———. 2004a. "Challenges and efforts to secure control systems," March. Washington, DC: General Accounting Office. http://www.gao.gov/cgi-bin/getrpt?

———. 2004b. "Critical infrastructure protection challenges and efforts to secure control systems," March. Washington, DC: General Accounting Office. March.http:www.gao.gov/cgi-bin/getrpt?GAO-04–354.

———. 2004c. "Nuclear regulation: NRC needs to more aggressively and comprehensively resolve issues related to the Davis-Besse nuclear power plant's shutdown," GAO-04–415, May 17. Washington, DC: General Accounting Office.

———. 2004d. "Nuclear security: DOE must address significant issues to meet the requirements of the new design basis threat," May 11. Washington, DC: General Accounting Office.

———. 2005. "Critical infrastructure protection: Challenges in addressing cybersecurity," July 19. Washington, DC: Government Accountability Office. http://www.gao.gov/cgi-bin/getrpt?GAO-05–827T.

———. 2006. "Homeland Security Department is taking steps to enhance security at chemical facilities, but additional authorityis needed," January. Washington, DC: Government Accountability Office. http://www.gao.gov/new.items/d06150.pdf.

Garrett, Thomas A., and Russell S. Sobel. 2002 "The political economy of FEMA disaster payments." St. Louis, MO: Federal Reserve Bank of St. Louis.

Geer, Daniel, et al. 2003. "CyberInsecurity: The cost of monopoly: How the dominance of Microsoft's products poses a risk to security," September 24. Washington, DC: Computer & Communicatons Industry Association. http://www .apple.com/education/technicalresources/pdf/cyberinsecurity.pdf, http://www .ccianet.org/papers/cyberinsecurity.pdf.

General Accounting Office. *See* GAO.

Gertz, B. 2002. "Nuclear plants targeted." *Washington Times,* January 31.

Gillett, Sharon Eisner, and Mitchell Kapor. 1997. "The self-governing Internet: Coordination by design." In *Coordination of the Internet,* edited by Brian Kahin and James Keller, 3–38. Boston: MIT Press.

Glaeser, Edward L. 2005. "Should the government rebuild New Orleans, or just give residents checks?" *Economists' Voice* 2: article 4.

Glasner, Joanna. 2003. "Want PC security? Diversify." *Wired,* September 25. http://www.wired.com/news/infostructure/0,1377,60579,00.html.

Glasser, Susan B., and Michael Grunwald. 2005. "Department's mission was undermined from the start." *Washington Post,* December 22.

Glenn, David. 2005. "Disaster sociologists study what went wrong in the response to the hurricanes, but will policy makers listen?" *Chronicle of Higher Education,* September 29. http://chronicle.com/free/2005/09/2005092904n .htm.

Gonzales, Vince. 2004. "Enron traders caught on tape." CBS, June 1. http://www .cbsnews.com/stories/2004/06/01/eveningnews/main620626.shtml.

Government Accountability Office. *See* GAO.

Grant, Don, and Andrew W. Jones. 2003. "Are subsidiaries more prone to pollute? New evidence from the EPA's Toxics Release Inventory." *Social Science Quarterly* 84 (March): 162–171.

Grant, Don Sherman II, Andrew W. Jones, and Albert J. Bergesen. 2002. "Organizational size and pollution: the case of the U.S. chemical industry." *American Sociological Review* 67 (June): 389–407.

Greenstein, Fred I. 2003. *The George W. Bush Presidency: An Early Assessment.* Baltimore, MD: Johns Hopkins University Press.

Grimaldi, James V., and Guy Gugliotta. 2001. "Chemical plants feared as targets." *Washington Post,* December 16.

Grunwald, Michael, and Susan B. Glasser. 2005. "Director who came to symbolize incompetence in Katrina predicted agency would fail." *Washington Post.* December 23.

Gup, Ted. 1994. "How FEMA learned to stop worrying about civilians and love the bomb." *Mother Jones,* January/February, 28–31+.

Hahn, Robert W., and Anne Layne-Farrar. 2006. "The law and economics of software security" (draft), March. Washington, DC: AEI-Brookings Joint Center for Regulatory Studies.

Hall, Mimi. 2006. "Brain drain hits Homeland Security." *USA Today,* March 29.

Hammond, Thomas. 2004. "Why is the intelligence community so difficult to redesign?" May 5. Paper presented at conference on Smart Practices toward In-

novation in Public Management, sponsored by the International Political Science Association Committee on the Structure and Organization of Government, University of British Columbia, Vancouver, June 15–17.

Hanley, Charles J. 2005. "Terrorists inept at waging war with chemicals." *Arizona Daily Star* (Phoenix), October 30.

Harris, Shane. 2002. "Tech insider: The homeland security market boom." Govexec.com (accessed January 9, 2005). http://www.govexec.com/dailyfed/020 102ti.htm.

———. 2006. "TIA lives on." *National Journal,* February 23.

Harrison, Bennett. 1994. *Lean and mean: The changing landscape of corporate power in the age of flexibility.* New York: Basic.

Harrison, K. 1999. "Talking with the donkey: Cooperative approaches to environmental protection." *Journal of Industrial Ecology* 2: 51–72.

Haygood, Wil, and Ann Scott Tyson. 2005. "A mob, crazy mentality." *Washington Post,* September 15.

Hedges, Stephen J., and Jeff Zeleny. 2001. "Mock terrorists breached security at weapons plants." *Chicago Tribune,* October 5.

Henry, Tom. 2004. "FirstEnergy No. 1 in Bush donations: Watchdogs say utilities sought favors." *Toledo (OH) Blade,* May 6.

Hirsh, Michael, and Michael Isikoff. 2002. "What went wrong." *Newsweek,* May 27, 28.

Hoffman, Andrew. 1997. *From heresy to dogma: An institutional history of corporate environmentalism.* San Francisco: New Lexington.

Holdeman, Eric. 2005. "Disasters keep coming but FEMA phased out." *New York Times,* August 31.

Homeland Security. 2005 "Oil pollution containment and recovery continues," September 18. Washington, DC: U.S. Coast Guard. http://www.piersystem .com/go/doc/425/104712/.

Hsu, Spencer S., and Steve Hendrix. 2005. "Hurricanes Katrina and Rita were like night and day." *Washington Post,* September 25.

Human Security Centre. *See* Centre, Human Security

Hunter, David. 2001. "Toulouse tragedy postscript." *Chemical Week,* October 10, 3.

IPCC. 2002. "Climate Change 2001: Impacts, Adaptation & Vulnerability." Geneva, Switzerland: Intergovernmental Panel on Climate Change.

ISDR. 2002. *Living with risk: A global review of disaster reduction initiatives.* Kobe, Japan: Inter-Agency Secretariat of the International Strategy for Disaster Reduction.

Jarocki, Martha Olson, and Pamela Calvert. 2004. "Chemical plant safety in the U.S.: Brought to you by the American Chemistry Council." *Global Pesticide Campaigner,* 14 December.

Jasper, James M. 1990. *Nuclear politics: Energy and the state in the United States, Sweden, and France.* Princeton, NJ: Princeton University Press.

Jenkins, William O. 2005. "Protection of chemical and water infrastructure". March. Washington, D.C: Government Accountability Office. http://www.gao .gov/cgi-bin/getrpt?GAO-05–327.

Johnson, Jeff. 1999. "Chemical accident data: Plethora of confusion." *Chemical and Engineering News,* March 15, 22–23.

———. 2004. "Plant security actions criticized." *Chemical & Engineering News,* November 29. http://pubs.acs.org/cen/staff/biojwj.html.

Josephson, D. H. 1994 "The great midwest flood of 1993." Silver Springs, MD: Department of Commerce, NOAA, National Weather Service.

Kady, Martin II. 2005. "House Republicans approve creation of permanent Homeland Security Committee." *CQ Today* (accessed January 5, 2005). http://CQ.com.

Kahn, Joseph. 2002. "Californians call Enron documents the smoking gun." *New York Times,* May 8.

Kahney, Leander. 2003. "Orrin Hatch, software pirate?" Wired.com (accessed June 10, 2005). http://www.wired.com/news/politics/1,59305–0.html.

Karasik, Theodore. 2002. *Toxic warfare.* Santa Monica, CA: Rand.

Kelzer, Gregg. 2006. "Spyware barely touches Firefox." TechWeb (accessed March 12, 2006). http://www.techweb.com/wire/security/179102616.

Kendra, James, Tricia Wachendorf, and Enrico L. Quarantelli. 2003. "The evacuation of Lower Manhattan by water transport on September 11: An unplanned success." *Joint Commission Journal on Quality and Safety,* 29 June, 316–318.

Kettl, Donald F. 2004. *Systems under stress: Homeland security and american politics.* Washington, DC: Congressional Quarterly Press.

Khurana, Rakesh. 2002. *Searching for a corporate savior: The irrational quest for charismatic CEOs.* Princeton, NJ: Princeton University Press.

King, Andrew A., and Michael J. Lenox. 2000. "Industry self-regulation without sanctions: The chemical industry's Responsible Care program." *Academy of Management Journal* 43: 698–716.

King, Robert P. 2006 "Dike poses danger to region." *Palm Beach (FL) Post,* May 28. http://www.palmbeachpost.com/storm/content/storm/reports/2006/lakeo_dike_overview.html.

Kleindorfer, Paul R., James C. Belke, Michael R. Elliott, Kiwan Lee, Robert A. Lowe, and Harold I. Feldman. 2003. "Accident epidemiology and the U.S. chemical industry: Accident history and worst-case data from RMP*Info." *Risk Analysis* 23 (October): 865–881. http://www.blackwell-synergy.com/links/doi/10.1111/1539–6924.00365/abs.

Klinenberg, Eric, and Thomas Frank. 2005. "Looting homeland security." *Rolling Stone,* December 29, 44–54.

Kocieniewski, David. 2005. "Facing the city, potential targets rely on a patchwork of security." *New York Times,* May 9.

———. 2006. "Despite 9/11 effect, rail yards are still vulnerable." *New York Times,* March 27.

Koff, Stephen. 2003. "Davis-Besse failure linked to new rules." *Cleveland Plain Dealer,* May 16.

Kolbert, Elizabeth. 2003. "Indian Point blank." *New Yorker,* March 3, 36–39.

Kosal, M. E. 2005. "Near term threats of chemical weapons terrorism." July 1. Paper presented at Center for Contemporary Conflict's WMD Proliferation Networks Conference, Naval Postgraduate School, Monterey, CA.

———. 2006. "Terrorism targeting industrial chemical facilities: Strategic motivations and the implications for US security." *Studies in Conflict and Terrorism,* in press.

Kosal, M. E., and D. E. Anderson. 2004. "An unaddressed issue of agricultural terrorism: A case study on feed security." *Journal of Animal Science* 82:3394–3400.

Krebs, Brian. 2006. "2005 Patch times for Firefox and Internet Explorer." *Washington Post,* February 15. http://blog.washingtonpost.com/securityfix/2006/02/2005_patch_times_for_firefox_a.html.

Kunkel, Kenneth E. 2003. "North American trends in extreme percipitation." *Natural Hazards* 29 (June): 291–305.

Kunreuther, Howard. 1998. "Introduction." In *Paying the price: The status and role of insurance against natural disasters in the United States,* edited by Howard Kunreuther and Richard J. Ross Jr., 1–16. Washington, DC: Joseph Henry.

Kuwabara, Ko. 2003. "Linux: a bazaar at the edge of chaos." *First Monday* (accessed March 3, 2003). http://firstmonday.org/issues/issue5_3/kuwabara/index.html.

LaMonica, Martin. 2005. "Massachusetts moves ahead sans Microsoft." CNET News (accessed March 12, 2006). http://news.com.com/2102-1012_3-5878869.html?tag=st.util.print.

Lane, Alexander. 2005. "Whitman: GOP foiled security efforts." *Newark Star-Ledger,* January 28.

La Porte, Todd R., and Paula M. Consolini. 1991. "Working in practice but not in theory: Theoretical challenges of high-reliability organizations." *Journal of Public Administration Research and Theory* 1:19–47.

Larson, Lee W. 1993. "The great midwest flood of 1993." Kansas City, MO: National Weather Service.

———. 1996. "The great USA flood of 1993." Washington, DC: U.S. Geological Survey. www.nws.noaa.gov/oh/hrl/papers/area/great.htm.

Lazerson, Mark. 1988. "Organizational growth of small firms: An outcome of markets and hierarchies?" *American Sociological Review* 53, no. 3:330–342.

Lazerson, Mark, and Gianni Lorenzoni. 1999. "The networks that feed industrial districts: A return to the Italian source." *Industrial and Corporate Change* 8:235–266.

Lean, Geoffrey. 2006. "Factory farms blamed for spread of bird flu." *Independent* (London), February 26.

Leisner, Pat. 2000. "Fla. militia leader sentenced." Associated Press, July 28. http://www.tmia.com/sentenced.html.

Leitenberg, Milton. 2004. *The problem of biological weapons.* Stockholm: Swedish National Defence College.

Leopold, Jason. 2005a. "Criminal trial related to California energy crisis may start soon." TruthOut (Accessed December 20, 2005). http://www.truthout.org/docs_2005/120105Z.shtml.

Leopold, Jason. 2005b. "FEMA chief Brown paid millions in false claims to help Bush win Florida votes." Free Press, September 15. http://www.freepress.org/departments/display/20/2005/1459.

Lerner, Eric J. 2003. "What's wrong with the electric grid?" *American Institute of Physics* (accessed December 29, 2005). http://www.aip.org/tip/INPHFA/vol-9/iss-5/p8.html.

Levitan, Marc. 2000. "Are chemical plants at risk from hurricane winds?" Department of Civil and Environmental Engineering, Louisiana State University (accessed June 14, 2005). http://hurricane.Isu.edu/_unzipped/levitan_paper1/levitan_paper1.PDF.

Lewis, Ted G. 2006. *Critical infrastructure protection in homeland security: Defending a networked nation.* New York: Wiley.

Lewis, Ted G., and Rudy Darken. 2005. "Potholes and detours in the road to critical infrastructure protection policy." *Homeland Security Affairs* 1: article 1.

Lieberman, Marvin B. 1987. "Market growth, economies of scale, and plant size in the chemical processing industries." *Journal of Industrial Economics* 36: 175–191.

Lindblom, Charles E. 1959. "The science of 'muddling through.'" *Public Administration Review* 19:79–88.

Lipton, Eric. 2004. "Big cities will get more in anti-terrorism grants." *New York Times,* December 22.

———. 2005. "Administration to seek antiterror rules for chemical plants." *New York Times,* June 15.

———. 2006a. "Audit finds mismanagement drained $1 billion project." *New York Times,* April 2.

———. 2006b. "'Breathtaking' waste and fraud in hurricane aid." *New York Times,* June 27.

———. 2006c. "Chertoff seeks a chemical security law, within limits." *New York Times,* March 22.

———. 2006d. "Come one, come all, join the terror target list." *New York Times,* July 12.

———. 2006e. "Disaster response: Watch TV, go home." *New York Times,* February 20.

———. 2006f. "FEMA calls, but top job is a tough sell." *New York Times,* April 2.

Lipton, Eric, and Scott Shane. 2005. "Leader of federal effort feels the heat." *New York Times,* September 3. http://www.nytimes.com/2005/09/03/national/nationalspecial/03fema.html.

Lochbaum, David. 2004. "U.S. nuclear plants in the 21st century: The risk of a lifetime," May. Washington, DC: Union of Concerned Scientists. www.ucsusa.org/clean_energy/nuclear_safety/page.cfm?pageID=1408.

Lohr, Steve. 2005. "Can this man reprogram Microsoft?" *New York Times,* December 11. www.nytimes.com/2005/12/11/business/yourmoney/11micro.html.

Lohr, Steve, and John Markoff. 2006. "Windows is so slow, but why?" *New York Times,* March 27.

Lynn, Barry. 2005. *End of the line: The rise and coming fall of the global corporation.* New York: Doubleday.

———. 2006. "Breaking the chain: The antitrust case against Wal-Mart." *Harpers,* July, 29ff.

MacAvoy, Paul W., and Jean W Rosenthal. 2004. *Corporate profit and nuclear*

*safety: Strategy at northeast utilities in the 1990s.* Princeton, NJ: Princeton University Press.

Mangels, John. 2003. "NRC officials who oversaw Davis-Besse are promoted." *Cleveland Plain Dealer,* July 9.

Mangels, John and John Funk. 2004. "Lid problem was on tape, but NRC didn't watch." *Cleveland Plain Dealer,* June 20.

Mansfield, Duncan. 2004. "U.S. nuclear plant cheated during drill." *Gulfport (MS) SunHerald,* January 27.

Markowitz, Gerald, and David Rosner. 2002. *Deceit and denial: The deadly politics of industrial pollution.* Berkeley: University of California Press.

McConnell, David. No date. "Mississippi River flood: 1993." University of Akron (accessed November 28 2005). http://enterprise.cc.uakron.edu/geology/natscigeo/Lectures/streams/Miss_Flood.pdf.

McGroddy, James C., and Herbert S. Lin. 2004. "A review of the FBI's Trilogy Information Technology modernization program." Washington, DC: National Research Council.

McKay, Betsy. 2005. "Katrina oil spill clouds future of battered suburb." *Wall Street Journal,* January 3.

McNeil, Donald G. Jr. 2006. "U.S. reduces testing for mad cow disease, citing few infections." *New York Times,* July 21.

McQuaid, John, and Mark Schleifstein. 2002. "Washing away." *New Orleans Times-Picayune* June 23 – 27. http://www.nola.com/hurricane/?/washingaway/.

McWilliams, Carey. 1949. *California: The great exception.* New York: Current.

Mednicoff, David M. 2005. "Compromising toward confusion: The 9/11 Commission report and American policy in the Middle East." *Contemporary Sociology* 34:107–115.

Meyer, Marshall W., and Lynne G. Zucker. 1989. *Permanently failing organizations.* Newbury Park, CA: Sage.

Milbank, Dana. 2004. "FBI budget squeezed after 9/11; request for new counterterror funds cut by two-thirds." *Washington Post,* March 22.

Mileti, Dennis S. 1999. *Disasters by design: A reassessment of natural hazards in the United States.* Washington, DC: Joseph Henry.

Miller, Matthew. 2000. "Microsoft's monopoly: Epic arrogance & mega-profits" *San Jose Mercury News,* May 2.

Mintz, John. 2003. "Homeland Security is struggling." *Washington Post,* September 7.

———. 2004. "DHS blamed for failure to combine watch lists." *Washington Post,* October 2.

Mohai, Paul, and Bunyan Bryant. 1992. "Environmental racism: Reviewing the evidence." In *Race and the incidence of environmental hazards,* edited by Paul Mohai, 163–176. Boulder, CO: Westview.

Molotch, Harvey L. 1970. "Oil in Santa Barbara and power in America." *Sociological Inquiry* 40:131–144.

Moss, David. 1999. "Courting disaster? The transformation of federal disaster policy since 1803." In *The Financing of catastrophe risk,* edited by Kenneth A Froot. 307–362. Chicago: University of Chicago Press.

Murphy, Dean E. 2005. "Storms put focus on other disasters in waiting." *New York Times,* November 15.

Murphy, Kevin. 2003. "Trouble grows at the Internet's root." *Computer Business Review Online,* October 20 (Accessed December 28, 2005). http://www.cbronline.com/latestnews/165c8acb5f79bb5780256dc50018bddd.

National Research Council. *See* Council, National Research.

Neumann, Peter G. 1995. *Computer related risks.* New York: ACM Press–Addison Wesley.

Noland, Bruce. 2005. "In storm, N.O. wants no one left behind." *New Orleans Times-Picayne,* July 24. http://delong.typepad.com/sdj/2005/09/new_orleanss_hu.html.

NRDC. 2003. "Chemical plant security: A tale of two Senate bills," September 9. Washington, DC: Natural Resources Defense Council. http://www.nrdc.org.

OpenSSL. 2005. "About the OpenSSL project." OpenSSL (accessed December 18, 2005). http://www.openssl.org/about/.

Oppel, Richard A. Jr. 2002. "How Enron got California to buy power it didn't need." *New York Times,* May 8.

Ornstein, Norman J. 2003. "Perspective on House reform of homeland security," May 19. Subcommittee on Rules, Select Committee on Homeland Security, U.S. House of Representatives. www.aei.org/news/newsID.1717514/news_detail.asp.

Outage, Task Force on. 2003. "Final report: Causes of the August 14th blackout in the United States and Canada: Causes and recommendations." US-Canada Power System Outage Task Force. Washington, DC: U.S. Department of Energy. https://reports.energy.gov/.

Pape, Robert. 2005. *Dying to win: The strategic logic of suicide terrorism.* New York: Random House.

Parachini, John V., Lynn E. Davis, and Timothy Liston. 2003. "Homeland security: A compendium of public and private organization's policy recommendations." Santa Monica, CA: Rand.

Pasternak, Douglas. 2001. "A nuclear nightmare: They look tough, but some plants are easy marks for terrorists." *U.S. News & World Report,* September 17, 44–46.

Perin, Constance. 2006. *Shouldering risks: The culture of control in the nuclear power industry.* Princeton, NJ: Princeton University Press.

Perrow, Charles. 1981. "Markets, hierarchy and hegemony: A critique of Chandler and Williamson." In *Perspectives on organizational design and behavior,* edited by Andrew Van de Ven and William Joyce, 371–386, 403–404. New York: Wiley Interscience.

———. 1986. "Economic theories of organizations." *Theory and Society* 15:11–45.

———. 1992. "Small firm networks." In *Networks and organizations,* edited by Nitin Nohria and Robert G. Eccles, 445–470. Boston: Harvard Business School Press.

———. 1999. *Normal accidents: Living with high risk technologies.* Princeton, NJ: Princeton University Press.

———. 2002. *Organizing America: Wealth, power, and the origins of corporate capitalism.* Princeton, NJ: Princeton University Press.

———. 2004. "Network-centric warfare: Some difficulties." In *Information assurance: Trends in vulnerabilities, threats, and technologies,* edited by Jacques S. Gansler and Hans Binnedijk, chap. 9. Washington, DC: National Defense University Press.

———. 2005. "Organizational or executive failure?" *Contemporary Sociology* 34 (March): 99–107.

———. 2006a. "The disaster after 9/11: The Department of Homeland Security and the intelligence reorganization." *Homeland Security Affairs* 2 (April). http://www.hsaj.org/hsa/volII/iss1/art3.

———. 2006b. "Disasters evermore? Reducing U.S. vulnerabilities." In *Handbook of disaster research,* edited by Havidan Rodriguez, Enrico L. Quarantelli, and Russell R. Dynes, 521–533. New York: Springer.

Peterson, Jonathan. 2005. "Grandma Millie, meet the detectives: An unlikely team unmasks Enron." *Los Angeles Times,* March 27.

Phillips, Zack. 2005. "Losing track of millions monthly, DHS agency explores accounting options." *CQ Homeland Security News,* August 5.

Pincus, Walter. 2002. "FEMA's influence may be cut under new department." *Washington Post,* July 24.

———. 2006. "Some lawmakers doubt DNI has taken intelligence reins." *Washington Post,* February 2.

Pinter, Nicholas. 2005. "One step forward, two steps back on U.S. floodplains." *Science* 308 (April 8): 207–208.

Piore, Michael, and Charles Sable. 1984. *The second industrial divide.* New York: Basic.

POGO. 2002. "Nuclear power plant security: Voices from inside the fence." Washington, DC: Project on Government Organization (accessed June 26, 2004). http://www.pogo.org/p/environment/eo-020901-nukepower.html.

Pooley, Eric. 1996. "Nuclear warriors" *Time,* March 4, 46–49.

———. 1997. "Nuclear safety fallout: A Time report shook up the NRC, which closed plants for repairs. But can it fix them." *Time,* March 17, 34.

Posner, Richard A. 2004. *Catastrophe: Risk and response.* Oxford and New York: Oxford University Press.

Powell, Walter W. 1990. "Neither markets nor hierarchy: Network forms of organization." *Research in Organizational Behavior* 12:295–336.

Powell, Walter W., Kenneth W. Koput, and Laurel Smith-Doerr. 1996. "Interorganizational collaboration and the locus of innovation: Networks of learning in biotechnology." *Administrative Science Quarterly* 41 (1):116–45.

Priman. No date. "The cost of power disturbances to industrial and digital economy companies." Palo Alto, CA: Electric Power Research Institute.

Prine, Carl. 2002. "Lax security exposes lethal chemical supplies" *Pittsburgh Tribune-Review,* April 7.

———. 2003. "Chemical plants still vulnerable." *Pittsburgh Tribune-Review,* November 16.

Purvis, Meghan, and Julia Bauler. 2004. "Irresponsible care: The failure of the chemical industry to protect the public from chemical accidents." New York: U.S. Public Interest Research Group Education Fund.

Quarantelli, Enrico L. 1954. "The nature and conditions of panic." *American Journal of Sociology* 60:267–275.

———. 2001. "Sociology of panic." In *International encyclopedia of the social and behavioral sciences,* edited by P.B. Baltes and N. Smelser. New York: Pergamon.

Rabinovitz, Jonathan. 1998. "A push for a new standard in running nuclear plants." *New York Times,* April 11.

Radsafe. 2004. "Radsafe archives: Search TEPCO." Vanderbilt University (Accessed June 25, 2004). www.vanderbilt.edu/radsafe/.

Reason, James. 1990. *Human error.* New York: Cambridge University Press.

Reynolds, Diana. 1990. "FEMA and the NSC: The rise of the national security state." *Covert Action Information Bulletin* 33 (Winter).

Rinaldi, Steven M., James P. Peerenboom, and Terence K. Kelly. 2001. "Identifying, understanding, and analyzing critical infrastructure interdependencies." *IEEE Control Systems* 21, no. 6 (December): 11–25.

Ripley, Amanda. 2004. "How safe are we?" *Time,* March 29, 34–36.

Roberts, Karlene H. 1993. *New challenges to understanding organizations.* New York: Macmillan.

Roberts, Patrick S. 2003. "Executive orders, congressional influence and the creation of the Department of Homeland Security." Paper presented at the American Political Science Association annual meeting, Philadelphia, August 28–31.

———. 2005. "Shifting priorities: Congressional incentives and the homeland security granting process." *Review of Policy Research* 22 (July): 437–449.

Rosenthal, Isadore. 2004. "Chemical Safety Board." *Risk Management Review* 4 (Winter): 6–7.

Ross, Brian. 2005. "Secret FBI report questions al Qaeda capabilities." ABC News, March 9.

Ross, Brian, and Rhonda Schwartz. 2004. "Official who criticized homeland security is out of a job." ABC News, December 21. http://abcnews.go.com/WNT.

Rudman, Warren B., Richard A. Clarke, and Jamie F. Metzl. 2003 "Emergency responders: Drastically underfunded, dangerously unprepared." Washington, DC: Council on Foreign Relations.

Sagan, Scott D. 1996. "When redundancy backfires: Why organizations try harder and fail more often." Paper presented at the American Political Science Association annual meeting, San Francisco, August 31.

———. 2003. "The problem of redundancy problem: will more nuclear security forces produce more nuclear security?" *Risk Analysis* 24 (4):935–46.

Sageman, Marc. 2004. *Understanding terror networks.* Philadelphia: University of Pennsylvania Press.

Saxenian, AnnaLee. 1996. "Regional advantage: Culture and competition in Silicon Valley and Route 128." Harvard University Press. Online book: http://name.umdl.umich.edu/HEB00993.

Scanlon, Joseph. 2001 "Emergent groups in established frameworks: Ottawa Carleton's response to the 1998 ice disaster." Article draft.

————. 2002. "Helping the other victims of September 11: Gander uses multiple EOCs to deal with 38 diverted flights." Ottawa: Emergency Communications Research Unit, Carleton University.

Schmid, Sonja. 2005. "Envisioning a technological state: Reactor design choices and political legitimacy in the Soviet Union and Russia." PhD dissertation, Department of Sociology and Technology Studies, Cornell University.

Schnaiberg, Allan. 1983. "Redistributive goals versus distributive politics: Social equity limits in environmental and appropriate technology movements." *Sociological Inquiry* 53:200–219.

Schroeder, Aaron D., Gary L. Wamsley, and Robert Ward. 2001. "The evolution of emergency management in America: From a painful past to a promising but uncertain future." In *Handbook of crisis and emergency management,* edited by Ali Farazmand, 357–419. New York: Marcel Dekker.

Scully, Malcolm G. 2002. "The natural world: Where land subsides into water." *Houston Chronicle,* April 26.

Shane, Scott, and Eric Lipton. 2005. "Stumbling storm-aid efforts put tons of ice on trips to nowhere." *New York Times,* October 2.

Shapiro, Jake. Forthcoming. "The greedy terrorist: A rational-choice perspective on terrorist organizations' inefficiencies and vulnerabilities." In *The political economy of terrorism finance and state responses: a comparative perspective,* edited by Jeanne K. Giraldo and Harold A. Trinkunas. Palo Alto, CA: Stanford University Press.

Shenon, Phillip. 2003a. "Antiterror money stalls in Congress." *New York Times,* February 13.

————. 2003b. "Former domestic security aides make a quick switch to lobbying." *New York Times,* April 29.

Shernkman, Rick. 2004. "Think again: Whatever happened to homeland security?" February 26. Washington, DC: Center for American Progress.

Showalter, P., and M. Myers. 1992. "Natural disasters as the cause of technological emergencies: A review of the decade 1980–1989." Boulder, CO: Natural Hazards Research and Applications Information Center, University of Colorado.

Shrader, Katherine. 2005. "Stay vigilant, anti-terror officials says." *Philadelphia Inquirer,* December 3.

Shrivastava, Paul. 1987. *Bhopal: Anatomy of a crisis.* Cambridge, MA: Ballinger.

Siemaszko, Andrew. 2003. "Complaint of Andrew Siemaszko against First Energy Nuclear Operating Company," February. Washington DC: U.S. Department of Labor, Occupational Safety and Health Administration.

Smith, Sandy. 2005. "Davis-Besse: A plan for change for a worse-case scenario?" *Occupational Hazard.* 67, no. 2 (February 18): 42–47.

Sniffen, Michael J. 2006. "FBI agent slams bosses at Moussaoui trial." Associated Press, March 20. http://apnews.myway.com/article/20060320/D8GFJ91O0.html.

Spaeth, Anthony. 2005. "China's toxic shock." *Time Asia,* http://www.time.com/time/asia/magazine/article/0,13673,501051205–1134807,00.html.

St. Clair, Jeffrey. 2004. "Bad days at Indian Point." Counterpunch.org, January 17–18. http://www.counterpunch.org/stclair01172004.html.

Staff. 1991a. "American notes: Nuclear power down for the count." *Time* August 26.

———. 1991b. "Phillips to pay $4 million for a fatal safety violations." *Atlanta Journal and Constitution,* August 23.

———. 2003. "Computer security experts say the recent 'SQL Slammer' worm, the worst in more than a year, is evidence that Microsoft's year-old security push is not working." CNN.com (accessed December 23, 2005). http://www.cnn.com/2003/TECH/biztech/02/01/microsoft.security.reut/index.html.

———. 2004. *The 9/11 Commission report.* New York: Norton.

———. 2005a. "Bush cronyism weakens government agencies." Bloomberg, September 30. www.bloomberg.com/apps/news?pid=10000087&sid=aJzwLcLRZiek.

———. 2005b. "FBI: Internet-based terrorism attacks unlikely." *Internet Security Review,* December. http://www.isr.net/isr/G/mnewsdec.html#FBI:%20Internet-Based.

———. 2005c. "FEMA: A legacy of waste." *South Florida Sun-Sentinel* (Miami), September 18, 19. http://www.sun-sentinel.com/news/local/southflorida/sfl-femareport,0,7651043.storygallery?coll=sfla-home-headlines.

———. 2005d. "'Hope is fading' at New Orleans Convention Center." NBC, MSNBC. http://www.truthout.org/docs_2005/090205I.shtml.

———. 2005e. "Judge stops Bush effort to eliminate federal workers' rights." *AFL-CIO News,* October 7. http://www.aflcio.org/joinaunion/ns10072005.cfm?RenderForPrint=1.

———. 2005f. "Katrina redux? Beaumont paper finds federal storm failures in Texas." *Editor & Publisher,* September 25.

———. 2005g. "Open source vs. Windows: Security debate rages on." *News-Factor Magazine* (accessed December 20). http://www.newsfactor.com/story.xhtml?story_id=102007ELB94U.

———. 2005h. "Time for chemical plant security." *New York Times,* December 27.

———. 2006a. "Ashcroft stakes out lucrative new ground." *Chicago Tribune,* January 13. http://columbiatribune.com/2006/Jan/20060113News005.asp.

———. 2006b. "Now what? The lessons of Katrina." *Popular Mechanics,* March. http://www.popularmechanics.com/science/earth/2315076.html?page=8&c=y.

Steingraber, Sandra. 2005. "The pirates of Illiopolis." *Orion,* May/June, 16–27. http://www.oriononline.org/pages/om/05–3om/Steingraber.html.

Stephenson, John B. 2005. "Federal and industry efforts are addressing security issues at chemical facilities, but additional action is needed," April 27. Washington, DC: Government Accountability Office.

Struck, Doug. 2005. "A flicker away from a blackout." *Washington Post,* June 21.

Swartz, Mimi, and Sherron Watkins. 2003. *Power failure: The inside story of the collapse of Enron.* New York: Doubleday.

Swiss Re. 2002. "Natural catastrophes and man-made disasters in 2001," January. Zurich: Swiss Re.

Swoboda, Frank. 1991. "Settlement set in \'90 plant blast." *Washington Post,* July 18.

Sylves, Richard, and William R. Cumming. 2004. "FEMA's path to Homeland Security: 1979–2003." *Journal of Homeland Security and Emergency Management* 1. http://www.bepress.com/jhsem/vol1/iss2/11.

TexPIRG. 2005. "BP facilities lead nation in chemical and refinery accidents since 1980, despite industry-touted safety measures," March 24. Austin, TX: TexPIRG.

Tidwell, Mike. 2005. "Goodbye, New Orleans: It's time we stopped pretending." *Orion* (accessed December 31, 2005). http://www.oriononline.org/pages/oo/sidebars/front/index_Tidwell.html.

Tierney, John 2005. "Going (down) by the book." *New York Times,* September 17.

Tierney, Kathleen. 2003. "Disaster beliefs and institutional interests: Recycling disaster myths in the aftermath of 9–11." In *Terrorism and disaster: New threats, new ideas: Research in social problems and public policy,* edited by Lee Clarke, 33–51. New York: Elsevier Science.

———. 2005. "The 9/11 Commission and disaster management: Little depth, less context, not much guidance." *Contemporary Sociology* 34 (March): 115–120.

Tierney, Kathleen, Christine Bevc, and Erica Kuligowski. 2006. "Metaphors matter: Disaster myths, media frames, and their consequences in Hurricane Katrina." *Annals, AAPSS* 604:57–81.

Tollan, Arne. 2002. "Land-use change and floods: What do we need most, research or management?" *Water Science and Technology* 45:183–190.

U.S. Congress. 2004. "Recommendations," September 30. House Select Committee on Homeland Security. http://hsc.house.gov/files/mini_report_sigs.pdf.

———. 2005. "Hurricane Katrina: The role of the Federal Emergency Management Agency," September 27. House Select Bipartisan Committee to Investigate the Preparations for and Response to Hurricane Katrina. http://katrina.house.gov/hearings/09_27_05/witness_list092705.html.

———. 2006. "Congressional preemption of state laws and regulations." June 6. Committee on Government Reform, minority staff, special investigations division. http://www.democrats.reform.house.gov/investigations.asp.

Uzzi, Brian. 1997a. "Social structure and competitioon in interfirm networks: The paradox of embeddedness." *Administrative Science Quarterly* 42 (March): 35–67.

———. 1996b. "The sources and consequences of embeddedness for the economic performance of organizations: The network effect." *American Sociological Review* 61 (August): 674–98.

Vadis, Michael A. 2004. "Trends in cyber vulnerabilities, threats, and countermeasures." In *Information assurance: Trends in vulnerabilities, threats, and*

*technologies,* edited by Jacques S. Gansler and Hans Binnendijk 99–114. Washington, DC: National Defense University.

Van Atta, Don Jr., 2002. "Bush's California energy stance faulted" *New York Times,* May 8.

van der Vink, G., et al. 1998. "Why the United States is becoming more vulnerable to natural disasters." *Eos, Transactions, American Geophysical Union* 79, no. 3 (November): 533, 537.

Van Natta, Don R., and David Johnston. 2002. "Wary of risk, slow to adapt, FBI stumbles in terror war." *New York Times,* June 2.

Vaughan, Diane. 1999. "The dark side of organizations: mistake, misconduct, and disaster." *Annual Review of Sociology* 25:271–305.

Verton, Dan. 2003. "DHS had little choice but to sign Microsoft deal, despite security flaws." Computerworld.com (accessed December 15, 2005). http://www .computerworld.com/governmenttopics/government/story/0,10801,83240,00 .html.

Wald, Matthew. 2001. "Nuclear sites ill-prepared for attacks, group says." *New York Times,* December 17.

———. 2002. "Nuclear plants said to be vulnerable to bombings." *New York Times,* March 20.

———. 2003a. "Group says test of nuclear plant's security was too easy." *New York Times,* September 16.

———. 2003b. "Ohio reactor's problems are said to persist." *New York Times,* May 4.

———. 2003c. "Regulators' wariness kept a damaged A-plant open." *New York Times,* January 4.

———. 2004. "Nuclear plant, closed after corrosion, will reopen." *New York Times,* March 9.

———. 2005a. "Experts assess deregulation as factor in '03 blackout." *New York Times,* September 16.

———. 2005b. "Study finds vulnerabilities in pools of spent nuclear fuel." *New York Times,* April 7.

Walker, David N. 2004. "The chief operating officer concept and its potential use as a strategy to improve management at the Department of Homeland Security." June 28. Washington, DC: U.S. General Accounting Office.

Wamsley, Gary L., and Aaron D. Schroeder. 1996. "Escalating in a quagmire: The changing dynamics of the emergency management policy subsystem." *Public Administration Review* 56 (May/June): 235–245.

Ward, Robert, Gary L. Wamsley, Aaron D. Schroeder, and David B. Robins. 2000. "Network organizational development in the public sector: A case study of the Federal Emergency Management Administration (FEMA)." *Journal of the American Society for Information Science* 51:1018–1032.

Warrick, Joby. 2006. "Custom-built pathogens raise bioterror fears." *Washington Post,* July 31.

Waugh, William L. Jr. 2000. *Living with Hazards, dealing with disasters.* Armonk, NY: M.E. Sharpe.

Wein, Lawrence M., and Yifan Liu. 2005. "Analyzing a bioterror attack on the food supply: The case of botulinum toxin in milk." *PNAS* 102 (July 12): 9984–9989. www.pnas.org_cgi_doi_10.1073_pnas.0408526102.

Welch, E. W., A. Mazur, and S. Bretschneider. 2000. "Voluntary behavior by electric utilities: Levels of adoption and contribution of the climate challenge program to the reductionof carbon dioxide." *Journal of Policy Analysis and Management* 19:407–425.

Whitlock, Craig. 2006. "Architect of the new war on the West." *Washington Post,* May 23.

Whitman, Christine Todd. 2005. *It's My Party Too.* New York: Penguin.

Wiener, Jon. 2005. "Cancer, chemicals and history." *Nation,* February 7. http://www.thenation.com/doc.mhtml?i=20050207&s=wiener.

Wildavksy, Aaron. 1988. *Searching for safety.* New Brunswick, NJ: Transaction.

Williamson, Oliver. 1975. *Markets and hierarchy: Analysis and antitrust implications.* New York: Free Press.

Wise, Charles R. 2002a. "Organizing for homeland security." *Public Administration Review* 62 (March/April): 131–144.

———. 2002b. "Reorganizing the federal government for homeland security: Congress attempts to create a new department." *Extensions,* Fall, 14–19.

Wise, Charles R., and Rania Nader. 2002. "Organizing the federal system for homeland security: Problems, issues, and dilemmas." *Public Administration Review* 62 (September): 44–57.

Wisner, Ben. 1999. "There are worse things than earthquakes: Hazard vulnerability and mitigation capacity in Greater Los Angeles." In *Crucibles of hazard: Megacities and disasters in transition,* edited by James K Mitchell. Tokyo and New York: United Nations University Press.

Yang, Catherine. 2005. "At stake: the net as we know it." *Business Week* (accessed December 26, 2005). http://proquest.umi.com/pqdweb?index=0&did=945043991&SrchMode=1&sid=1&Fmt=3&VInst=PROD&VType=PQD&RQT=309&VName=PQD&TS=1135399848&clientId=13766#fulltext.

Zegart, Amy B. 1999. *Flawed by design: The evolution of the CIA, JCS, and NSC.* Stanford, CA: Stanford University Press.

———. 2007. *Intelligence in wonderland: 9/11 and the organizational roots of failure.* Princeton, NJ: Princeton University Press.

# Index

------------------

9/11 attacks: chemical facilities security after, 193–99, 204; counterterrorism efforts as response to, 8–9, 60–61, 68–71, 75–76, 77, 81–82, 126, 311, 322; diverted air traffic during, 214; first responders and communication during, 49; nuclear facilities security after, 135, 138–39, 176; Patriot Act as response to, 102; planning and execution of, 305–9, 313; preparedness prior to, 293–94. *See also* 9/11 Commission Report

9/11 Commission Report: civil liberties provisions in, 125; on communication failures between intelligence agencies, 79; reorganization as response to, 107, 122, 308; on terrorist organizations, 305–6, 308

Afghanistan, 76, 105, 305–7

*Against All Enemies* (Clarke), 80–81, 105–6

agency theory, 331

agriculture: in California Central Valley, 33–34; crop insurance, 45, 46–47; deconcentration of, 16–17, 323; flooding risks increased by modern, 17–18; food safety and concentration of pro-

duction, 7, 16–17, 299–300, 323; water availability and, 33–34; wetlands polluted by, 20

airline industry: computer network failures at airports, 258; deregulation of, 7; diverted air traffic during 9/11 response, 214; security measures at airports, 117–18; as terrorist target, 7, 70, 126

Allbaugh, Joseph, 96, 109, 243

"all hazards" approach: DHS and, 127; FEMA and, 50, 54; reduction of target size as effective, 310, 314

Al Qaeda: chemical attacks and, 180–81; electrical grid as target for planned cyber attack by, 215; intelligence regarding threat, 73, 75, 79, 311; Iraq invasion and, 105, 305; as network organization, 304–5; nuclear plants as targets of, 135; organizational failures in, 305–7, 308–9; strategic objectives of, 69–70, 313. *See also* 9/11 attacks

Alverez, Robert, 136

American Chemistry Council (ACC), 190, 197, 203

American Society of Chemical Engineers (ASCE), 181–82